Climate Variability Handbook

Climate Variability Handbook

Edited by **Andrew Hyman**

New York

Published by Callisto Reference,
106 Park Avenue, Suite 200,
New York, NY 10016, USA
www.callistoreference.com

Climate Variability Handbook
Edited by Andrew Hyman

International Standard Book Number: 978-1-63239-116-2 (Hardback)

Printed in the United States of America.

Contents

Preface

An elucidative study on climate variability has been provided in this book. With the increasing awareness about the adversities due to climatic change, interest is developing in the people to understand various aspects of climatic variability and its impact at regional levels. For those interested in climatic change management, it provides detailed information regarding the dynamics of climatic variability and will prove to be a valuable resource. It discusses innovative methods to supervise climatic changes, evaluating climate risks and forecasting the consequences, analysis of present status of the knowledge on the subject and available evidences. This book is easy to understand and easily accessible to practitioners, researchers, students and policy makers.

This book has been the outcome of endless efforts put in by authors and researchers on various issues and topics within the field. The book is a comprehensive collection of significant researches that are addressed in a variety of chapters. It will surely enhance the knowledge of the field among readers across the globe.

It is indeed an immense pleasure to thank our researchers and authors for their efforts to submit their piece of writing before the deadlines. Finally in the end, I would like to thank my family and colleagues who have been a great source of inspiration and support.

Editor

Section 1

West African Monsoon in State-of-the-Science Regional Climate Models

M. B. Sylla, I. Diallo and J. S. Pal

Additional information is available at the end of the chapter

1. Introduction

The western Sahel (e.g. rectangle in Figure 1) climate is dominated by the West African Monsoon (WAM), a large-scale circulation characterized by reversal in direction of winds in the lower levels of the atmosphere from the Atlantic Ocean transporting moisture into land. Although it occurs only during a small portion of the annual cycle between May and September, the WAM is a climatological feature of major social importance to local populations over West Africa whose economy relies primarily on agriculture. In response to several decades of below normal rainfall experienced since the late 1960s, numerous studies have identified various factors that control the monsoon variability. Among them are the variability of ocean Sea Surface Temperatures (SSTs) (e.g. Fontaine et al. 1998), continental land surface conditions (e.g. Wang and Eltahir 2000) and atmospheric circulation (e.g. Nicholson and Grist 2001; Jenkins et al. 2005).

Within the atmospheric circulation pattern, are embedded a number of rainfall producing systems (e.g. Figure 2). One of the most prominent features is the African Easterly Jet (AEJ), a mid-tropospheric (600–700 mb) core of strong zonal winds (up to ~10 m s^{-1}) that travels from East to West Africa. The disturbances around this zonal circulation, the African Easterly Waves (AEWs), have been identified as key driver of convection and rainfall patterns (Diedhiou et al. 1998). Most of the convective rainfall follows the south-north-south displacement of the Intertropical Convergence Zone (ITCZ) with a mean upward motion that reaches 200 mb. At this level, the Tropical Easterly Jet (TEJ), associated with the Asian monsoon outflow, circulates across West Africa during the boreal summer season. These features form the well-defined meridional structure of WAM circulation related to the mean summer monsoon rainfall (e.g. Figure 3).

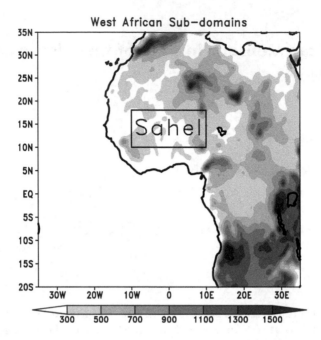

Figure 1. West Africa AMMA domain and topography highlighting the Western Sahel region.

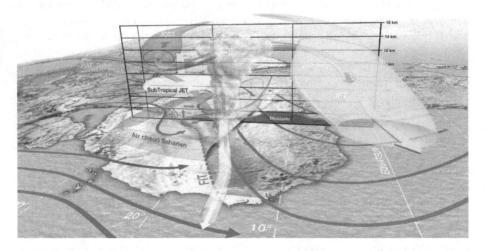

Figure 2. schematic view of the West African Monsoon System adopted from Lafore et al. (2010). FIT stands for ITD (InterTropical Discontinuity), "Air chaud Saharien" stands for 'Warm Saharian Air", JEA stands for AEJ (African Easterly Jet), JET stands for TEJ (Tropical Easterly Jet), "Air sec" stands for "Dry Air".

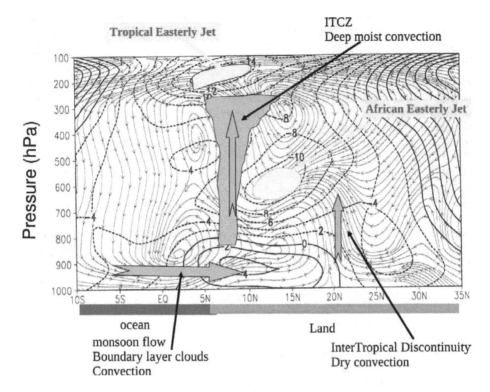

Figure 3. Mean Meridional Circulation (stream lines) and associated mean Zonal Wind (m/s in contours) over West Africa during the summer season adopted from Hourdin et al. (2010).

The intensity and position of the above features not only influence the amount of rainfall but also its variability, indicating a strong scale interaction between the different elements of the WAM (Redelsperger et al. 2002). For example over the Sahel, a more equatorward position of the AEJ often corresponds with drought conditions (Jenkins et al. 2005; Nicholson 2008) while a stronger than average TEJ often results in wetter climates. It is important to note that not all displacement of the AEJ is associated with wetter/drier Sahel. Precipitation in the region also appears to be strongly related to SSTs (e.g. Giannini et al. 2003; Lu and Delworth 2005; Biasutti et al. 2008). Furthermore, active AEWs are mainly generated by significant occurrence of deep moist and shallow cumulus convection (Hsieh and Cook 2008) and are thus typically expected during wetter periods.

The total seasonal rainfall is the result of successive convective events, which are either local or organized as Mesoscale Convective Systems (MCSs) structured along the AEJ where the interaction between the tropical waves and convection plays the dominant role at both synoptic and intra-seasonal time scales (e.g. Mathon and Laurent 2001; Gaye et al. 2005; Mohr and Thorncroft, 2006). The onset of the rainy season over the Sahel is partly dictated by the

Saharan Heat Low (SHL) (Hagos and Cook 2007; Thorncroft et al. 2011) which plays a significant role in the monsoon jump process, an abrupt shift of the rainbelt characterized by a northward extension of deep convective rainfall from the coast (about 6N) between May-June to about 10-12N in July-August.

Modeling this complex interplay between the monsoon dynamical features is of primary importance for an accurate representation of rainfall over the Sahel and thus paramount for a better understanding of the West African climate response to global warming. Such modeling simulations have often been performed using Global Climate Models (GCMs). However, the GCMs often have problems in representing accurately the main WAM features, presumably due to coarse grid spacing typically used (Hourdin et al. 2010; Sylla et al. 2010a; Xue et al. 2010). Studying the West African climate using Regional Climate Models (RCMs), which are typically applied at a high resolution than GCMs, has therefore become an area of active research in the last decades. The added resolution allows for a better representation of fine-scale forcing and land surface heterogeneity such as vegetation variations, complex topography and coastlines, which are important aspects of the physical response governing the local and regional climate change signal (e.g. Paeth et al. 2005; Rummukainen 2010; Sylla et al. 2012a). The overall aim of this chapter is to assess the extent at which the state-of-the-science RCMs are able to simulate the mean climatology, mean annual cycle and inter-annual variability of rainfall over the western Sahel as well as the relationship of rainfall to the monsoon dynamical structures at different time-scales. This is critical to better assess the key processes associated with the interactions between the atmosphere, land and ocean, as well as between the dynamics and convection. The use of RCMs can contribute to improving our understanding of, and ultimately our ability to predict rainfall variability. Section 2 reviews previous applications of RCMs over West Africa. Section 3 presents our results of mean climatology, annual cycle and inter-annual variability of rainfall over the Sahel along with its interactions with the different WAM features from state-of-the-science RCMs. Section 4 summarizes the conclusions and provides an outlook of future directions of RCMs research over West Africa.

2. Regional climate models applications over West Africa

RCMs use a limited area grid driven at the lateral boundaries from GCM output or reanalysis data. SSTs are generally prescribed, except when an ocean model is enabled. RCMs are employed to add additional detail to GCM simulations or to conduct process studies requiring additional fine-scale detail over what GCMs capture (Giorgi and Mearns 1999). Examples of process studies include investigation of the impacts of land cover change, dust and biomass burning, and climate change.

Over the past 15 years, RCMs have been extensively used as dynamical downscaling tool for different applications over West Africa (e.g. Vizy and Cook 2002; Gallée et al. 2004; Paeth et al. 2005; Afiesimama et al. 2006; Kamga and Buscarlet 2006; Pal et al. 2007; Hagos and Cook 2007; Sylla et al. 2010b; Pohl and Douville 2011; Murthi et al. 2011; Diallo et al. 2012a).In general, these studies have shown that RCMs can adequately represent the WAM climatology and its variability.

GCMs, integrated at typical grid spacing (100s km) are unable to represent key features of the WAM such as AEWs, AEJ and TEJ. As mentioned above, one of the primary purposes for using an RCM is to enhance or add-value to individual and multi-model RCM ensembles with respect to their driving GCM or reanalysis field. Generally speaking, RCMs do not correct gross biases in the GCM or reanalysis fields. In other words, garbage in equals garbage out. Sylla et al. (2009), however, concluded that the systematic errors in output from their RCM simulations were driven only to a minor extent from the errors in the driving large-scale GCM fields and thus, were tied to the representation of local/regional processes. In addition, Druyan et al. (2010) found that their individual model errors varied considerably in space and from model to model while Paeth et al. (2011) demonstrated that their multi-model ensemble clearly exhibited added value in WAM rainfall with respect to the European Reanalysis-Interim (ERA-Interim; Dee et al. 2011) driving field. Diallo et al. (2012b) suggested that the multi-model RCM ensembles should be based on different driving GCMs for improved performance.

Recent studies have proposed ways to further validate RCMs dynamics over West Africa using process-based evaluation. Such evaluation examines how well a particular phenomenon, process or feedback is represented in an RCM. Adequate RCM performance in simulating the monsoon dynamical features variability is crucial to reasonably represent rainfall over the Sahel (Diallo et al. 2012c). Performance can also be related to the domain size and location of lateral boundaries (Browne and Sylla 2012). It is important to emphasize that a key feature in this process-based evaluation is to relate the ability of an RCM to represent the process in future climate and a realistic simulation of climate change (e.g. Paeth et al. 2009; Sylla et al. 2010a; Patricola and Cook 2010 Mariotti et al. 2011; Skinner et al. 2012; Abiodun et al. 2012). For example, Sylla et al. (2010c) found that simulations using the International Centre for Theoretical Physics Regional Climate Model (RegCM3; Pal et al. 2007) reproduced the observed relationship between Sahel rainfall and WAM features, such as the AEJ, TEJ and AEWs at both the intra-seasonal and inter-annual time-scales, demonstrating that the model performance was of sufficient quality for climate process and mechanism studies over West Africa.

There is substantial evidence that the representation of the land surface and land-atmosphere coupling is particularly important for the WAM regions (e.g. Patricola and Cook 2008; Steiner et al. 2009, van den Hurk and van Meijgaard 2010). For instance, Abiodun et al. (2010), using RegCM3 conducted land-surface sensitivity experiments and found that deforestation increases the intensity of the AEJ core, thus reducing the northward transport of moisture needed for rainfall over West Africa. Steiner et al. (2009) illustrated that the use of Community Land Model version 3 (CLM3; (Oleson et al. 2004) instead of Biosphere Atmosphere Transfer Scheme (BATS; Dickinson et al. 1993) as land-surface model in RegCM3 induced a weaker temperature gradient which shifted the AEJ southward, in line with observations, and greatly improved the wet model bias. The key role played by vegetation feedbacks in future climate predictions over West Africa has also been highlighted by Wang and Alo (2012). They found that dynamic vegetation feedback reverses the predicted future trend, leading to a substantial increase of annual rainfall. The response of the WAM to soil moisture anomalies, planetary boundary layer and Southern Sudan bioclimatic zone soil types was reported by Moufouma-Okia and Rowell (2010), Flaounas et al. (2011) and Zaroug et al. (2012) respectively. Their results demonstrated that changes in these surface parameters significantly impact the WAM precipitation.

Of particular interest is the application of RCMs to improve our understanding of the WAM dynamics and features. Using Mesoscale Model version 5 (MM5; Grell et al. 1994), Vizy and Cook (2002) studied the response of the WAM to tropical oceanic heating. Other RCM studies explored the relationship between tropical or Mediterranean SST and the WAM (Messager et al. 2004; Gaetani et al. 2010; Fontaine et al. 2010). They identified the main physical mechanisms connecting the WAM precipitation pattern with Gulf of Guinea, northern Atlantic and Mediterranean SSTs. The dynamics of the observed abrupt latitudinal shift of maximum precipitation from the Guinean coast into the Sahel region around May-June, the so called "monsoon jump", has also been investigated in various studies applying different RCMs (Ramel et al. 2006; Sijikumar et al. 2006; Hagos and Cook 2007; Drobinski et al. 2009; Sylla et al. 2010c). These studies highlight the key role played by SHL, diabatic heating and AEJ in this process. Hsieh and Cook (2005) examined the relationship between the ITCZ, the AEJ and the AEWs and recently, Sylla et al. (2011) addressed the role of the representation of deep convection on key elements of the West African summer monsoon climate using RegCM3. Interestingly, the former found that AEWs develop more readily in a simulation with a weak AEJ and a strong ITCZ pointing out the importance of cumulus convection and the associated release of latent heat while the latter showed that the deep convection scheme is of lesser importance for the genesis and growth of AEWs but is a key factor for the simulation of a more realistic AEJ and for the west coast wave development.

It is evident that RCMs are valuable tools for understanding WAM dynamics and the interactions between its different components. Most of the studies mentioned above, however, are performed using a single RCM, which are known to incorporate unpredictable and large random errors that substantially influence the simulation at shorter time scales (Vanvyve et al. 2008). Therefore, it is not surprising that in the last few years, significant efforts have been employed to establish coordinated frameworks using several RCMs aimed at improving the characterization of the WAM at various time-scales. Such frameworks include the West African Monsoon Modelling and Evaluation (WAMME; Xue et al. 2010; Druyan et al. 2010), African Multidisciplinary Monsoon Analysis (AMMA; Ruti et al. 2010), the Ensemble-based Predictions of Climate Change and their Impacts (ENSEMBLES; Paeth et al. 2011) and recently the Coordinated Regional Downscaling Experiment (CORDEX; Giorgi et al. 2009; Jones et al. 2011).

The AMMA-MIP (AMMA Model Intercomparison Project) has been coordinated to evaluate GCMs and RCMs in terms of their ability to reproduce the mean West African climate and, in particular, the seasonal and intra-seasonal variations of rainfall and associated dynamical structures for two contrasting years: 2000 and 2003 (Hourdin et al. 2010). More recently, multiyear (1989-2007) simulations of the West African climate over the AMMA domain (e.g.

Figure 1) have been completed by eight state-of-the-science RCMs forced by the newly developed ERA-Interim reanalysis at the boundaries for a more robust assessment. To date, this constitutes the most comprehensive database incorporating all the relevant variables for a better characterization of the WAM. Table 1 summarizes the different RCMs and their respective physical schemes and latest references.

RCMs	GKSS-CCLM	METNO-HIRHAM	ICTP-REGCM3	KNMI-RACMO2.2b	SMHI-RCA	METO-HC_HadRM3.0	UCLM-PROMES	MPI-M-REMO
Institute	GKSS Forschungszentrum Geesthacht GmbH (GKSS), Germany	The Norvegian Meteorological Institute, Norway	Abdus Salam International Centre for theoretical Physics, Italy	Koninklijk Nederlands Meteorologisch Instituut, Netherlands	Swedish Meteorological and hydrological institute, Sweden	Met Office-Hadley Centre, UK	Universidad de Castilla La Mancha (UCLM), Spain	Max Planck Institute, Germany
Horizontal	0.44° (50 km)	0.44° (50 km)	50 km	0.44° (50 km)	0.44° (50 km)	0.44° (50 km)	50 km	0.44° (50km)
Convective scheme	Tiedtke (1989)	Tiedtke (1989), Nordeng (1994)	Grell (1993); Fritch and Chappell (1980)	Tiedtke (1989)	Kain and Fritsch (1990)	Gregory and Rowntree (1990), Gregory and Allen (1991)	Kain and Fritsch (1990)	Tiedke (1989)
Radiation scheme	Ritter and Geleyn (1992)	Morcrette (1991), Giorgetta and Wild (1995)	Kiehl et al. (1996)	Fouquart and Bonnel (1980)	Savijarvi (1990), Sass et al. (1994)	Edwards and Slingo (1996)	Anthes et al. (1987), Garand (1983)	Morcrette et al. (1986) Giorgetta and Wild (1995)
Land surface scheme	TERRA3D, BATS, Grasselt et al. (2008)	Dümenil and Todini (1992)	BATSE1E, Dickinson et al. (1993)	TESSEL, Van der Hurk et al. (2000)	RCA land surface model, Samuelsson et al. (2006)	MOSES2 Essery et al. (2003)	SECHIBA, Ducoudré et al. (1993)	Hangemann (2002) Rechid et al. (2009)
Cloud microphysics scheme	Kessler (1969), Lin et al. (1983), Ritter and Geleyn (1992)	Sundquist (1978)	SUBEX, Pal et al. (2000)	Tiedtke (1993)	Rasch and Kristjànsson (1998)	Smith (1990), Jones et al. (1995)	Hsie et al. (1984)	Lohmann and Roeckner (1996)
Reference	Rockel et al. (2008)	Christensen et al. (2006)	Pal et al. (2007)	Lendrinck et al. (2003)	Kjellstróm et al. 2005	Jones et al. (2004)	Sahchez et al 2004	Jacob et al. (2001)

Table 1. Summary the different Regional Climate Models and their respective main physical schemes and latest references

3. Evaluation of state-of-the-Science RCMs over West Africa

In this section, we analyse and intercompare the performance of a set of eight (8) RCMs in simulating the mean climatology, annual cycle and interannual variability of rainfall over West Africa during the monsoon season and the related atmospheric features modulating this variability. The models are driven at the lateral boundaries by the ERA-Interim reanalysis, the most recent and improved reanalysis product available. These simulations are performed as part of the AMMA/ENSEMBLES multi-model intercomparison project - Research Theme 3 (RT3).

3.1. Mean summer monsoon climatology

Before evaluating and comparing the mean annual cycle and inter-annual variability of rainfall and its interaction with the main WAM features, it is important to evaluate the mean summer monsoon climatology over West Africa. The temperature field from observations Climate Research Unit (CRU; Mitchell et al. 2004) and University of Delaware (UDEL; Legates and Willmott 1990); Figure 4a, b), the ERA-Interim reanalysis (Figure 4c), the eight RCMs (Figure 4d-k) and their ensemble mean (Figure 4l) depict a zonal pattern with cooler temperatures along the Gulf of Guinea, increasing northward to reach a maximum around the area of the SHL centered at 25N. This area is well-defined by the lower pressure system and higher temperature values observed and simulated there. It is worth noting that the minima are found over the orographic peaks of Guinea Highlands, Jos Plateau and Cameroon Mountains, demonstrating the capability of the RCMs to capture the fine-scale features over regions of steep topography. A cold bias of about 2 degrees Celsius prevails over the Gulf of Guinea and this feature is common for ERA-Interim (the driving fields of the RCMs) and for all the RCMs, except MPI-REMO. However, MPI-REMO, along with METNO-HIRHAM and MET-O-HadRM3P, exhibit a warm bias over the SHL. As a result, the multi-model ensemble fails to outperform individual RCM members and the driving field, highlighting the importance of the individual model in the performance of the ensemble mean. It is difficult to unambiguously determine the causes of the RCMs temperature biases as they depend on a number of factors, including cloudiness, surface albedo, temperature advection and surface water and energy fluxes (Sylla et al. 2012b). It should be emphasized that none of the experiments incorporate aerosol effects and the inclusion of dust radiative forcing would likely reduce the bias and improve the simulation of surface air temperature and resulting effects on the WAM circulation (Konare et al. 2008; Camara et al. 2010; Solmon et al. 2012). Considering these uncertainties, and considering that typical RCM biases for seasonal surface temperature are within the range of 2 degrees Celsius (e.g. Jones et al. 1995; McGregor et al. 1998; Hudson and Jones 2002; Konare et al. 2008; Hernandez-Diaz et al. 2012; Giorgi et al. 2012), these state-of-the-science RCMs bias are in line with other RCMs applications over other regions.

Concerning the summer monsoon rainfall climatology, CRU (Figure 5a), Global Precipitation Climatology Project (GPCP; Adler et al. 2003) (Figure 5b), ERA-Interim (Figure 5c) and the superimposed wind at 850 mb are compared to the corresponding RCM fields (Figure 5d-k) and their ensemble mean (Figure 5l). The main summer rainfall is positioned in a zo-

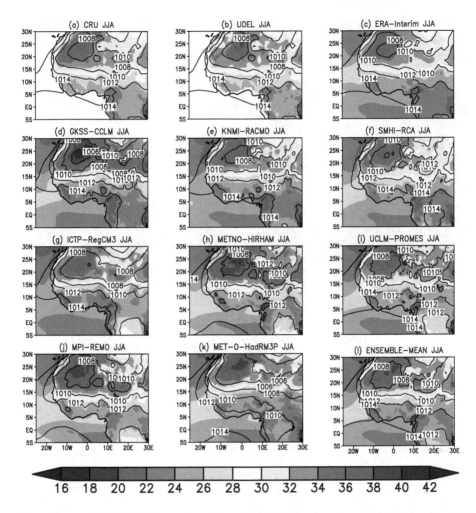

Figure 4. Averaged 1990-2007 JJA 2-meter Temperature (degree Celsius) and superimposed Mean Sea Level Pressure (MSLP, in hPa) in contour from: (a) CRU observation and ERA-Interim MSLP, (b) UDEL observation and ERA-Interim MSLP, (c) ERA-Interim reanalysis, (d) GKSS-CCLM, (e) KNMI-RACMO, (f) SMHI-RCA, (g) ICTP-RegCM3, (h) METNO-HIR-HAM, (i) UCLM-PROMES, (j) MPI-REMO, (k) MET-O-HadRM3P and (l) the RCMs ensemble mean.

nal and tilted band between 5N and 15N, with rainfall decreasing to the north and south, as shown in the plots of the observations (CRU and GPCP) and reanalysis (ERA-Interim). Precipitation maxima are, however, located in orographic regions such as Guinea Highlands, Jos plateau, and Cameroon Mountains which also experience the coldest temperatures indicating large evaporative cooling over these areas (e.g. Figure 4). This precipitation pattern is associated with moist southwesterly winds from the Atlantic Ocean, the so-called moisture

laden monsoon flow. Key differences across the observations are that CRU shows a discontinuity in the band of maximum rainfall over West Africa and much lower intensities over Jos
Plateau and that GPCP has lower rainfall amounts along the coastlines of Cameroon/Nigeria
highlands. Uncertainty in rainfall observations are a key factor preventing a rigorous and
unambiguous evaluation of RCMs over the region (Sylla et al. 2012c). The use of multiple
observed rainfall products can, however, help quantify the uncertainty.

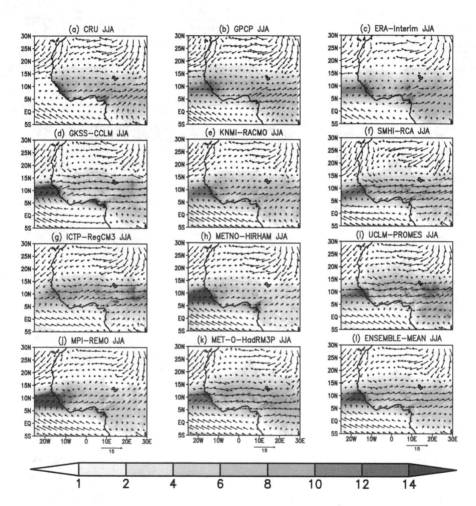

Figure 5. Averaged 1990-2007 JJA Precipitation (mm/day) and superimposed Wind Arrows from: (a) CRU observation
and ERA-Interim Wind, (b) GPCP observation and ERA-Interim Wind, (c) ERA-Interim reanalysis, (d) GKSS-CCLM, (e)
KNMI-RACMO, (f) SMHI-RCA, (g) ICTP-RegCM3, (h) METNO-HIRHAM, (i) UCLM-PROMES, (j) MPI-REMO, (k) MET-O-
HadRM3P and (l) the RCMs ensemble mean.

The RCMs display a similar spatial pattern in both rainfall and low-level winds compared to the observations and the ERA-Interim reanalysis but exhibit different bias magnitudes suggesting that the differences between the RCMs mainly arise from their internal dynamics and physics. Contrary to surface temperature, the ensemble mean of precipitation generally outperforms the the individual RCMs and the driving ERA-Interim data, illustrating the advantages of multi-model assessments of the WAM rainfall (Paeth et al. 2011, Nikulin et al. 2012; Diallo et al. 2012a). Most of the discrepancies occur along the ITCZ indicating that the RCMs substantially differ in their ability to simulate the interactions between the WAM elements and the deep convection.

To better understand this issue, an assessment of how well the RCMs reproduce the mean annual cycle, inter-annual rainfall variability and their relationship with the different dynamical features (AEJ, TEJ, AEWs and SHL) is performed in the next section. It should be noted that the various RCMs well capture their mean position with different magnitudes compared to the ERA-Interim driving field (not shown for brevity).

3.2. Mean annual cycle

The mean annual cycle of the WAM rainfall is presented in Figure 6 through a time-latitude Hovmoller diagram. It corresponds to a south-north-south displacement of the ITCZ, which is characterized by successive active and break phases of the convective activity. This meridional cross-section analysis averaged from 10W to 10E provides a good framework to assess RCM skill in simulating mean annual cycle and intra-seasonal variations of the WAM and associated mechanisms responsible for the rainfall variability (Hourdin et al. 2010).

CRU and GPCP observations as well as ERA-Interim reanalysis clearly identify the three distinct phases of the annual cycle: the initiation or onset phase (March-May), the high rain period (June-August), and the southward retreat of the rainbelt (September-October) as documented by Le Barbe et al. (2002). The onset period is characterized by a northward extension of the rainbelt from the coast to about 4N. An abrupt shift, the monsoon jump (Sultan and Janicot 2003), occurs at the beginning of June in both observation datasets and reanalysis, when the rain core moves rapidly northward to about 10N. This is the beginning of the high rain season in Sahel region and the sudden termination of heavy precipitation along the Guinea Coast. In September, a sharp southward retreat of the rainfall belt occurs, corresponding to the last phase of the WAM season. The AEJ corresponds closely to the rainfall variations and undergoes a poleward migration that peaks in August over northern Sahel around 13N, as depicted by the ERA-Interim reanalysis (Figure 6c). An interesting feature is that the rainfall core is located below the AEJ, suggesting that the easterly shear tends to favor deep convection south of the jet axis in line with previous studies (Thorncroft and Blackburn 1999; Diongue et al. 2002; Mohr and Thorncroft 2006; Fontaine et al. 2010). Unlike the AEJ, the TEJ appears only during the boreal summer season (June-September) with its core located around 5N. The TEJ promotes mid-level convergence by establishing upper-level divergence, indicating that the ITCZ oscillations during the annual cycle occur in association with the deep convective ascent bounded by the jets axes and levels (Nicholson et al. 2008; Sylla et al. 2010b).

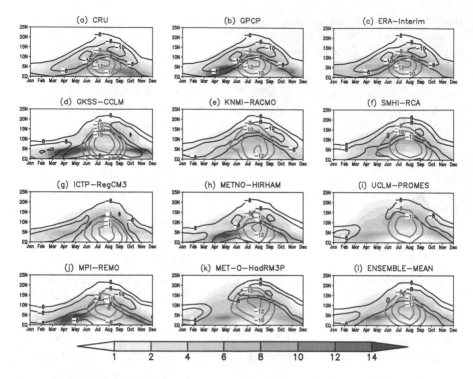

Figure 6. Time-Latitudes Hovmoller diagram of averaged 1990-2007 monthly Precipitation (mm/day) and superimposed 700 hPa (black contours) and 200 hPa (blue contours) Zonal Wind (m/s) from: (a) CRU observation and ERA-Interim Wind, (b) GPCP observation and ERA-Interim Wind, (c) ERA-Interim reanalysis, (d) GKSS-CCLM, (e) KNMI-RACMO, (f) SMHI-RCA, (g) ICTP-RegCM3, (h) METNO-HIRHAM, (i) UCLM-PROMES, (j) MPI-REMO, (k) MET-O-HadRM3P and (l) the RCMs ensemble mean.

Moving to the comparison with the RCMs, it is evident that the three distinct phases of the mean annual cycle, i.e. the monsoon jump, the poleward migration of the AEJ and the appearance and strengthening of the TEJ, are well reproduced. A number of differences can be observed among the experiments with regard to the magnitude and spatial extent of the features. For instance, GKSS-CCLM, METNO-HIRHAM and MPI-REMO overestimate the pre-monsoon rainfall, which is likely due to an early appearance of strong easterlies that would favor intense convection to the south of them. It is important to note that the RCMs exhibit different sensitivity in terms of their response to the intensity of the WAM elements. For example, GKSS-CCLM overestimates the TEJ while UCLM-PROMES and MET-O-HadRM3P underestimate the AEJ during the pre-monsoon period. This, once again, points to the importance of the different internal dynamics and physics of the models. Ensemble means can help to compensate for these errors, improve simulation results, and instigate a more clear connection between the WAM rainfall and atmospheric characteristics at the intra-seasonal time-scale (e.g. Figure 6l).

The SHL and its relationship to the rainfall annual cycle is presented as a time-latitude Hov-moller diagram of surface air temperature for CRU, UDEL, ERA-Interim and the superim-posed mean sea level pressure, the different RCMs and their ensemble mean (Figure 7a-l). The SHL appears as the area of very low pressure and higher temperature. The RCMs show good agreement in reproducing the intensification and northward migration of the SHL from the northern Sahel in March-April to the Sahara in July-August. This drives a progressive in-crease of lower-level temperature gradient between the Gulf of Guinea and the Sahara, strengthening and shifting toward the north of the AEJ, which ultimately favors intense con-vection in the latitudes below. Notable overestimations of about 2 to 4 degrees Celsius appear during the pre-monsoon and the monsoon periods in GKSS-CCLM, METNO-HIRHAM and MPI-REMO consistent with the early appearance and stronger easterlies discussed previous-ly. Unexpectedly this warm bias in the Sahara during the peak monsoon period does not trig-ger a stronger and more poleward AEJ, nor a wider rainfall band indicating that the land surface-atmosphere coupling may be less responsive in some RCMs.

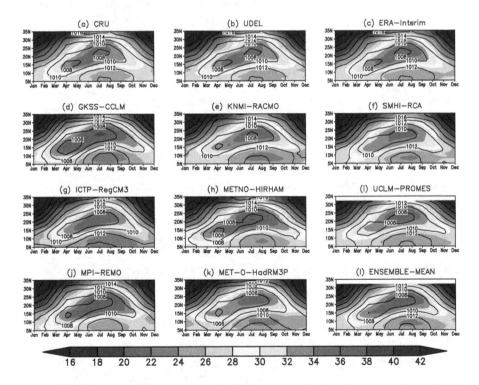

Figure 7. Time-Latitudes Hovmoller diagram of averaged 1990-2007 monthly Temperature (mm/day) and superim-posed Mean Sea Level Pressure (MSLP, in hPa) in contour from: (a) CRU observation and ERA-Interim Wind, (b) UDEL ob-servation and ERA-Interim Wind, (c) ERA-Interim reanalysis, (d) GKSS-CCLM, (e) KNMI-RACMO, (f) SMHI-RCA, (g) ICTP-RegCM3, (h) METNO-HIRHAM, (i) UCLM-PROMES, (j) MPI-REMO, (k) MET-O-HadRM3P and (l) the RCMs ensemble mean.

3.3. Inter-annual variability

Figure 8 shows observed and simulated mean June, July, and August (JJA) rainfall anomalies over the Sahel (e.g. see outlined box in Figure1). As a reference, we use the CRU data, which exhibits similar variability to the UDEL and GPCP data with a significant (at 95%) Pearson's correlation coefficient of more than 0.9 (Figure 8a and b). The anomalies are calculated with respect to the precipitation mean derived from the full 18-year period 1990–2007. The area averages of precipitation anomalies are normalized by the standard deviation derived from the 1990–2007 time series. The first interesting aspect to note is that all the RCMs outperform the ERA-Interim reanalysis in terms of the coefficient of correlation. This suggests that all of the RCMs produce a clear added value to the driving ERA-Interim data. This also implies that the inter-annual rainfall variability across the Sahel is less dependent on the large-scale boundary conditions and that the regional forcing, such as higher resolution land–surface interactions with synoptic processes may play important role. Another aspect is among the models, only GKSS-CCLM and MET-O-HadRM3P fail to capture the interannual variation. The ensemble mean of the RCMs not only reproduces the overall variability but also most of

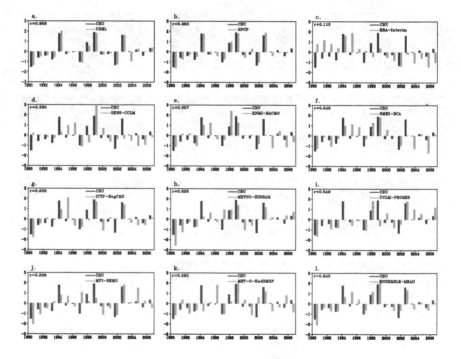

Figure 8. Interannual Variability of Standardized Seasonal (JJA) Precipitation Anomalies from 1990 to 2007 for: (a) CRU and UDEL observations, (b) CRU and GPCP observations (c) ERA-Interim reanalysis and CRU, (d) GKSS-CCLM and CRU, (e) KNMI-RACMO and CRU, (f) SMHI-RCA and CRU, (g) ICTP-RegCM3 and CRU, (h) METNO-HIRHAM and CRU, (i) UCLM-PROMES and CRU, (j) MPI-REMO and CRU, (k) MET-O-HadRM3P and CRU, and (l) the RCMs ensemble mean of the anomalies and CRU.

the direction and magnitude of the individual anomalies with by far the highest correlation coefficient of 0.84 (significant at 95%). This supports the use of RCMs and their multi-model ensembles to explore drivers of inter-annual variability over the Sahel.

To further investigate this issue, the four wettest years (1994, 1998, 1999, and 2003) and four driest years (1990, 1993, 1997, and 2002) are examined to assess the ability of the RCMs to simulate the characteristic circulation of these contrasting regimes. The difference between the dry and wet years for the 700 and 200 hPa zonal winds are respectively shown in Figures

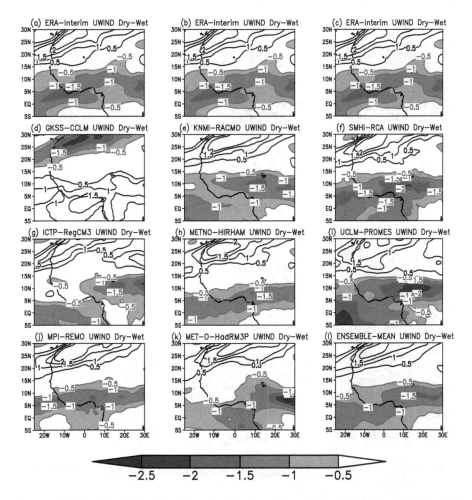

Figure 9. Difference of Dry (1990, 1993, 1997, and 2002) minus Wet Years (1994, 1998, 1999, and 2003) for the JJA 700 hPa Zonal Wind (m/s) for: (a,b,c) ERA-Interim reanalysis, (d) GKSS-CCLM, (e) KNMI-RACMO, (f) SMHI-RCA, (g) ICTP-RegCM3, (h) METNO-HIRHAM, (i) UCLM-PROMES, (j) MPI-REMO, (k) MET-O-HadRM3P and (l) the RCMs ensemble mean.

9 and 10. The ERA-Interim and all the RCMs, except GKSS-CCLM, exhibit an increase (decrease) in the 700 hPa zonal wind speed south (north) of 15°N during dry years. This indicates that the AEJ undergoes a southward displacement and thus shifts the deep convection equatorward, consistent with a drier Sahel. Compared to the individual RCM members, the ensemble mean produces a pattern most similar to that of the ERA-Interim. At the upper levels (200 hPa), only the ICTP-RegCM3 and UCLM-PROMES are able to depict a clear weakening of the TEJ seen in the ERA-Interim data implying a weaker upper-level divergence and hence much less lower-level convergence consistent with the rainfall deficit.

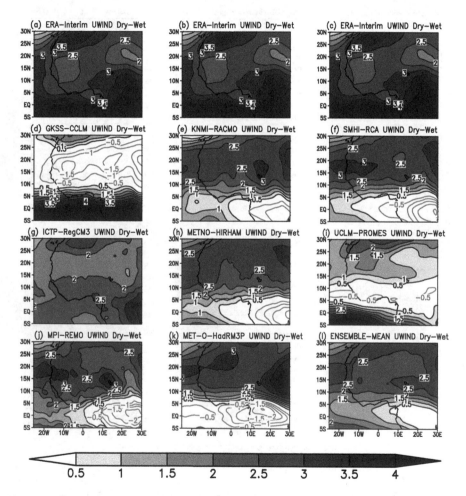

Figure 10. Difference of Dry (1990, 1993, 1997, and 2002) minus Wet Years (1994, 1998, 1999, and 2003) for the JJA 200 hPa Zonal Wind (m/s) for: (a,b,c) ERA-Interim reanalysis, (d) GKSS-CCLM, (e) KNMI-RACMO, (f) SMHI-RCA, (g) ICTP-RegCM3, (h) METNO-HIRHAM, (i) UCLM-PROMES, (j) MPI-REMO, (k) MET-O-HadRM3P and (l) the RCMs ensemble mean.

The corresponding composite for surface air temperature and superimposed mean sea level pressure for observations (CRU and UDEL), the ERA-Interim reanalysis, the RCMs and their ensemble mean are displayed in Figure 11. The observations and the reanalysis portray a dipole pattern with decreased temperature in Western Sahara accompanied by higher mean sea level pressure, and an increase in the Eastern Sahara along with a drop of the mean sea level pressure system. This illustrates a reduction of the SHL intensity and a possible shift of its core to the east during dry years over the Sahel. It is worth noting that the temperature pattern does not show any changes over the Gulf of Guinea with the CRU and UDEL observations. This implies substantial weakening of the meridional temperature gradient between the Gulf of Guinea and the Sahara thus preventing a northward displacement of the AEJ consistent with a drier Sahel. Furthermore, the observations and reanalysis data exhibit elevated temperature values over the Sahel probably due to strong soil moisture - temperature feedback. For the most part, all of the RCMs represent the weakening of the SHL in Western Sahara and the shift of its core to the East. The complete observed pattern, however, is fully replicated by SMHI-RCA and MPI-REMO and to some extent by KNMI-RACMO, UCLM-PROMES, GKSS-CCLM and the ensemble mean. Some of the RCMs, for instance ICTP-RegCM3, METNO-HIRHAM and MET-O-HadRM3P simulate cooler temperatures over some regions of the Sahel indicating weaker soil moisture - temperature feedback.

3.4. African Easterly Waves

AEWs are key drivers of climate variability in West Africa during the monsoon season. They are defined as disturbances around the zonal circulation of the mid-tropospheric AEJ and have been identified as the main mechanism organizing convection and rainfall patterns in this region (Diedhiou et al., 1999). They are triggered by localized finite-amplitude perturbations associated with latent heating upstream of the region of growth (Thorncroft et al. 2008; Leroux and Hall 2009) and maintained by combined baroclinic and barotropic conversions around the AEJ (Hsieh and Cook 2007). Hence the key role of deep convection in initiating the AEWs over the Darfur and Ethiopian highlands, and favoring its subsequent development is a consolidated result (Berry and Thorncroft 2005; Mekonnen et al. 2006; Sylla et al. 2011). Therefore, the existence of different convection schemes in climate models (e.g. Table 1) is likely to drive large uncertainties inherent to the simulation of AEWs (Ruti and Dell'Aquila, 2010). To gain insight into these uncertainties, we evaluate AEW activity simulated by different RCMs with similar and different convection schemes.

For this, we adopt two different bandpass filters in the 700 hPa daily meridional winds to separate the 3- to 5-day and the 6- to 9-day wave regimes as observed by Diedhiou et al. (1998) and Hsieh and Cook (2005). The wave activity is then obtained by averaging the seasonal mean variance of the filtered 700 hPa daily meridional winds for the period 1990-2008. This is shown for ERA-Interim (that drives the RCMs), National Center for Environmental Prediction (NCEP; Kalnay et al. 1996), and each of the RCMs and their ensemble mean respectively in Figure 12 The daily meridional winds are not available for KNMI-RACMO and UCLM-PROMES, and thus are not analyzed in this section.

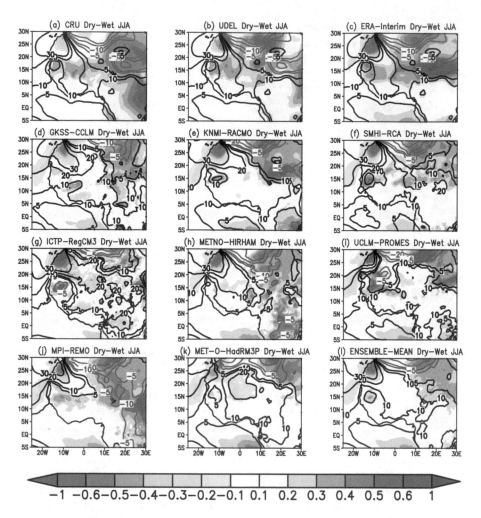

Figure 11. Difference of Dry (1990, 1993, 1997, and 2002) minus Wet Years (1994, 1998, 1999, and 2003) for the JJA 2-meter Temperature (degree Celsius) and superimposed Mean Sea Level Pressure (MSLP, in hPa) for: (a) CRU observation and ERA-Interim Wind, (b) UDEL observation and ERA-Interim Wind, (c) ERA-Interim reanalysis, (d) GKSS-CCLM, (e) KNMI-RACMO, (f) SMHI-RCA, (g) ICTP-RegCM3, (h) METNO-HIRHAM, (i) UCLM-PROMES, (j) MPI-REMO, (k) MET-O-HadRM3P and (l) the RCMs ensemble mean.

Striking differences are observed between the reanalysis. The ERA-Interim reanalysis depicts the higher 3- to 5-day wave activity slightly north of the ITCZ (around 15N) and off the west coast between 5N and 25N, and some weaker activity around the Gulf of Guinea and southern Sudan bioclimatic zone while NCEP reanalysis shifts the inland wave activity further north (around 20N). These differences between reanalysis products in terms of the rep-

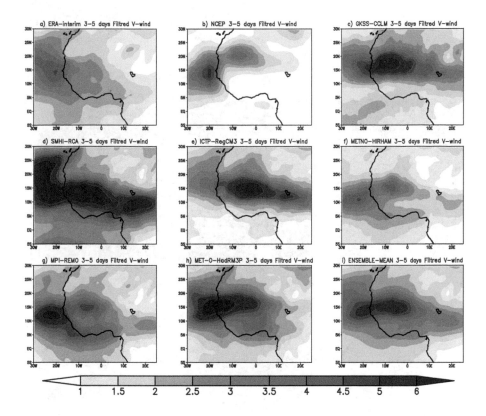

Figure 12. Mean (1990-2007) Variance of the 3- to 5-day filtered JJA 700 hPa Meridional Wind from (a) ERA-Interim reanalysis, (b) NCEP reanalysis (c) GKSS-CCLM, (d) SMHI-RCA, (e) ICTP-RegCM3, (f) METNO-HIRHAM, (g) MPI-REMO, (h) MET-O-HadRM3P and (i) the RCMs ensemble mean.

resentation of AEW activity have been identified and analyzed in detail by Ruti and Dell'Aquila (2010). The RCMs and their ensemble simulate a similar pattern to ERA-Interim but considerably overestimate the 3- to 5-day wave activity over the Sahel. This implies that these overestimates originate internally within the RCMs although some precursors of AEWs may enter the domain through the boundary forcing.

Compared to the 3- to 5-day, the core of the 6- to 9-day wave activity (Figure 13a-i) is generally weaker, probably due to their intermittent nature (Diedhiou et al. 1998; Hsieh and Cook 2007; Sylla et al. 2011), and located further north in both reanalysis datasets and all of the RCMs. Some of the largest model overestimates in both wave regimes, for instance SMHI-RCA and ICTP-RegCM3, are associated with excessive rainfall along the ITCZ, highlighting the key role of convection in triggering and maintaining the AEWs in these models. It is worth noting that GKSS-CCLM also exhibits stronger activity for both wave regimes but lower rainfall amounts implying the existence of dry convection fueling the AEWs. Contrary to the 3- to 5-day wave regime, the ensemble mean of the 6- to 9-day activity substantially improves the individual RCM members; however, much of this improvement results from cancellation of errors of opposite signs.

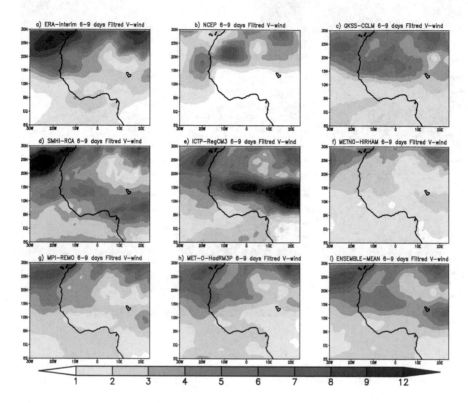

Figure 13. Mean (1990-2007) Variance of the 6- to 9-day filtered JJA 700 hPa Meridional Wind from (a) ERA-Interim reanalysis, (b) NCEP reanalysis (c) GKSS-CCLM, (d) SMHI-RCA, (e) ICTP-RegCM3, (f) METNO-HIRHAM, (g) MPI-REMO, (h) MET-O-HadRM3P and (i) the RCMs ensemble mean.

It is thus evident that the different RCMs, incorporating different convection schemes, produce quite different representation of AEWs regimes. Whether this is going to affect their dynamics during contrasting (wet and dry) years needs to be explored. We consider the same previous set of wet and dry years and examine the ability of the RCMs to reproduce the AEW composite pattern (Figure 14). The ERA-Interim data exhibit a reduction of the 3- to 5-day wave activity along a tilted zonal band extending from Cameroon Mountains to the Atlantic Ocean and encompassing the Gulf of Guinea and the Sahel region south of 15N (Figure 14a). This is consistent with less frequent convection events and thus a drier Sahel. The lower activity off the west coast is a robust result as is also observed in the NCEP data (Figure 14b). This pattern portrayed by ERA-Interim is well replicated by the ensemble mean, MPI-REMO and MET-O-HadRM3P (although they overestimate the magnitude of the difference) and to some extent by ICTP-RegCM3. The 6- to 9-day wave regime composite does not show any significant change over the Sahel (not shown).

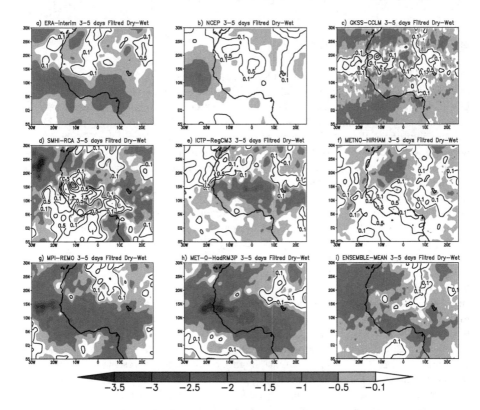

Figure 14. Difference of Dry (1990, 1993, 1997, and 2002) minus Wet Years (1994, 1998, 1999, and 2003) for the Mean Variance of the 3- to 5-day filtered JJA 700 hPa Meridional Wind from (a) ERA-Interim reanalysis, (b) NCEP reanalysis (c) GKSS-CCLM, (d) SMHI-RCA, (e) ICTP-RegCM3, (f) METNO-HIRHAM, (g) MPI-REMO, (h) MET-O-HadRM3P and (i) the RCMs ensemble mean.

A number of considerations can be highlighted based on the results of interannual variability along with the differences between the dry and wet years pertaining to the WAM circulation features. First, the monsoon circulation represented by the ERA-Interim data is not related to the inter-annual rainfall variability. This may be understandable as rainfall in this product is mostly simulated (Dee et al. 2011). Second, GKSS-CCLM fails to reproduce the inter-annual variability partly because of its inability to capture the circulation change characteristics of dry/wet periods over the Sahel (mostly AEJ and AEWs). Third, the interannual variability is more consistent with the change in AEJ and AEWs rather than in the TEJ for the majority of the models, suggesting that these rainfall variations are less sensitive to the upper-level divergence. Therefore, while almost all RCMs successfully replicate the inter-annual variability of precipitation over the Sahel, only few of them (ICTP-RegCM3, UCLM-PROMES and MPI-REMO) along with the multi-model ensemble mean are able to capture most of the related monsoon circulation patterns during contrasting years.

4. Conclusion and outlook

This chapter provided a review of recent RCMs applications for the WAM and its variability at a wide range of spatial and time scales. These applications illustrate the increased interest in research activities concerning the WAM that have been substantially and consistently developed during the last decade. Such interest has been motivated in large measure by the vulnerability of the West African countries to recurrent drought episodes. The socio-economic repercussions of drought have prompted the scientific community to improve seasonal forecasts to be able to accurately predict onset dates and growing season length among other relevant parameters related to agriculture. Understanding the WAM is thus crucial for developing early warning of imminent drought and mitigation strategies. We highlight the key role played by RCMs in improving our understanding of the nature of the interactions between the WAM rainfall and the features triggering and maintaining it. Much of the progress achieved can be attributable to the AMMA project (Redelsperger et al. 2006) through AMMA/ENSEMBLES, which provided a comprehensive dataset to study the multiple scales interactions, the related processes and the uncertainties that characterize the simulation of the WAM. We then analyze and compare eight state-of-the-science RCMs in simulating these interactions, identify key biases and seek to provide useful information for future modeling work. In particular, we establish that a good representation of the variability of the relevant atmospheric circulation features is paramount for the simulation of the monsoon precipitation pattern at different time-scales. Overall the multi-model ensemble and the majority of the RCMs provide encouraging results in the representation of the WAM rainfall climatology, the mean annual cycle and Sahel interannual variability along with their connection to the different atmospheric features during the summer monsoon season and contrasting wet and dry years. RCMs are thus suitable tools (albeit not perfect) for investigating the dynamics of the main features triggering and maintaining the WAM precipitation.

However, we should emphasize that an important limitation about RCMs is that the model solution depends strongly on boundary conditions provided by reanalyses or GCMs. Thus, the quality of the regional simulations relies on observations supplying the boundary conditions or the GCMs performance. In addition, although RCMs can be run at higher resolution (10s of km), unresolved scales are always present and thus RCMs still depend on parameterization quality. Another deficiency may be related to the domain size and the location of the lateral boundaries, i.e. whether the RCMs simulate or acquire from the boundaries the circulation features influencing precipitation over West Africa. Such shortcomings induce large uncertainties for both present-day and projections of future climate, suggesting merit in performing multi-model ensemble of RCM simulations nested within multiple GCMs.

Since AMMA, a new and broader framework, the Coordinated Downscaling Experiment (CORDEX; Giorgi et al. 2009, Jones et al. 2011) recently established by the World Climate Research Program (WCRP) is underway. CORDEX intends to foster an international coordinated effort to produce improved multi-model RCMs-based high resolution climate change scenarios in such a way that the uncertainty can be minimized, quantified and effectively communicated to the end-users for informed decision making. For West Africa, the program offers an unprecedented opportunity to advance knowledge of the African monsoons response to anthropogenic climate change. However, the datasets generated need to be comprehensive enough to allow a full characterization of key processes in order to provide new insights about the future evolution of the coupled atmosphere-land-ocean monsoon systems.

Acknowledgements

Mouhamadou Bamba Sylla is supported by the National Science Foundation (NSF) through Grant number: 1049186. Therefore, we would like to express our gratitude to the project's PI Guiling Wang (University of Connecticut). We also would like to thank gratefully the AMMA/ENSEMBLES modeling groups for sharing these datasets.

Author details

M. B. Sylla[1], I. Diallo[2] and J. S. Pal[1]

1 Loyola Marymount University, Seaver College of Science and Engineering, Department of Civil Engineering and Environmental Science, Los Angeles, CA, USA

2 Laboratory for Atmospheric and Ocean Physics - Simeon Fongang, Polytechnic School, University Cheikh Anta Diop, Dakar, Senegal

References

[1] Abiodun, B. J, Pal, J, Afiesimama, E. A, Gutowski, W. J, & Adedoyin, A. (2010). Modelling the impacts of deforestation on Monsoon Rainfall in West Africa. IOP Conf. Series and Environmental Science 13, doi: , 1755-1315.

[2] Abiodun, B. J, Adeyema, D. Z, Oguntunde, P. G, Salami, A. T, & Ajayi, V. O. (2012). Modeling the impact of reforestation on future climate in West Africa. Theoretical and Applied Climatology, doi:s00704-012-0614-1.

[3] Adler, R. F, et al. (2003). The version-2 Global Precipitation Climatology Project (GPCP) monthly precipitation analysis (1979-present). Journal of Hydrometeor , 4(6), 1147-1167.

[4] Afiesimama, A. E, Pal, J. S, Abiodun, B. J, Gutowski, W. J, & Adedoyin, A. (2006). Simulation of West African monsoon using the RegCM3. Part I: Model validation and interannual variability. Theoretical and Applied Climatology , 86, 23-37.

[5] Anthes, R. A, Hsie, E. Y, & Kuo, Y. H. (1987). Description of the Penn State/NCAR Mesoscale Model Version 4 (MM4). NCAR Technical Note-282 NCAR, Boulder, CO 80307.

[6] Berry, G. J, & Thorncroft, C. (2005). Case Study of an Intense African Easterly Wave. Monthly. Weather Review , 133, 752-766.

[7] Biasutti, M, Held, I. M, Sobel, A. H, & Giannini, A. (2008). SST forcings and Sahel rainfall variability in simulations of the twentieth and twenty first centuries. Journal of Climate , 21, 3471-3486.

[8] Browne NAKSylla MB. 2012. Regional climate model sensitivity to domain size for the simulation of the West African monsoon rainfall. International Journal of Geophysics, (2012). Article ID 625831, doi:

[9] Camara, M, Jenkins, G. S, & Konaré, A. (2010). Impacts of dust on West African climate during 2005 and 2006. Atmospheric Chemistry Physic Discuss , 10, 3053-3086.

[10] Christensen, O. B, Drews, M, Christensen, J. H, Dethloff, K, Ketelsen, K, Hebestadt, I, & Rinke, A. (2006). The HIRHAM Regional Climate Model. Version 5. DMI technical report 06-17: Available online at http://www.dmi.dk/dmi/tr pdf.

[11] Dee and co-authors (2011). The ERA-Interim reanalysis: configuration and performance of the data assimilation system. Quarterly Journal of the Royal Meteorology Society doi:qj.828., 137, 553-597.

[12] Diallo, I, Sylla, M. B, Gaye, A. T, & Camara, M. (2012a). Intercomparaison de la climatologie et de la variabilité interannuelle de la pluie simulée au Sahel par les modèles climatiques regionaux.

[13] Diallo, I, Sylla, M. B, Giorgi, F, Gaye, A. T, Camara, M, & Multi-model, b. GCM-RCM ensemble based projections of temperature and precipitation over West Africa for the early 21st century. International Journal of Geophysics, (2012). Article ID 972896, doi:

[14] Diallo, I, Sylla, M. B, Camara, M, & Gaye, A. T. (2012c). Interannual variability of rainfall and circulation features over the Sahel based on multiple regional climate models simulations. Theoretical and Applied Climatology. (In Press).

[15] Dickinson, R. E, Henderson, S. A, & Kennedy, P. J. (1993). Biosphere-Atmosphere Transfer Scheme (BATS) version 1E as coupled to the NCAR Community Climate Model. NCAR Tech. rep. TN-387+STR, 72 pp.

[16] Diedhiou, A, Janicot, S, Viltard, A, & De Felice, P. (1998). Evidence of two regimes of easterly waves over West Africa and tropical Atlantic. Geophysical Research Letter , 25, 2805-2808.

[17] Diedhiou, A, Janicot, S, Viltard, A, De Felice, P, & Laurent, H. (1999). Easterly wave regimes and associated convection over West Africa and tropical Atlantic: results from NCEP/NCAR and ECMWF reanalyses. Climate Dynamics , 15, 795-822.

[18] Diongue, A, Lafore, J. P, Redelsperger, J. L, & Rocca, R. (2002). Numerical study of a Sahelian synoptic weather system: initiation and mature stages of convection and its interactions with the large scale dynamics. Quarterly Journal of the Royal Meteorology Society , 128, 1899-2007.

[19] Drobinski, P, Bastin, S, Janicot, S, Dabas, A, Delville, P, Reitebuch, O, & Sultan, B. On the Late Northward Propagation of the West African Monsoon in Summer (2006). in the Region of Niger/Mali. Journal of Geophysical Research 114, D09108, doi:JD011159.

[20] Druyan, L. M, Feng, J, Cook, K. H, Xue, Y, Fulakeza, M, Hagos, S. M, Konare, A, Moufouma-okia, W, Rowell, D. P, Vizy, E. K, & Ibrah, S. S. (2010). The WAMME regional model Intercomparison study. Climate Dynamics doi:s00382-009-0676-7., 35, 175-192.

[21] Ducoudré, N, Kaval, K, & Perrier, A. (1993). Sechiba, a new set of parameterizations of the hydrologic exchanges at the land-atmosphere interface within the LMD atmospheric general circulation model. Journal of Climate , 6, 248-273.

[22] Dümenil, L, & Todini, E. (1992). A rainfall runoff scheme for use in the Hamburg climate model, in Advances in Theoretical Hydrology, A Tribute to James Dooge, European Geophysical Society Series on Hydrological Sciences, edited by J. P. O'Kane, , 1, 129-157.

[23] Edwards, J. M, & Slingo, A. (1996). Studies with a flexible new radiation code. I: choosing a configuration for a large-scale model. Quarterly Journal of the Royal Meteorology Society , 122, 689-719.

[24] Essery RLHBest MJ, Betts RA, Cox PM. (2003). Explicit representation of subgrid heterogeneity in a GCM land surface scheme. Journal of Hydrometeorology , 4, 530-543.

[25] Flaounas, E, Bastin, S, & Janicot, S. (2011). Regional climate modelling of the 2006 West African monsoon: sensitivity to convection and planetary boundary layer parameterisation using WRF. Climate Dynamics doi:s00382-010-0785-3., 36, 1083-1105.

[26] Fontaine, B, Garcia-serrano, J, Roucou, P, Rodriguez-fonseca, B, Losada, T, Chauvin, F, Gervois, S, Sijikumar, S, Ruti, P, & Janicot, S. Impacts of Warm and Cold situations in the Mediterranean Basins on the West African monsoon: observed connection patterns (1979-(2006). and climate simulations. Climate Dynamics dois00382-009-0599., 35, 95-114.

[27] Fontaine, B, Trazaska, S, & Janicot, S. (1998). Evolution of the relationship between near and global and Atlantic SST modes and the rainy season in West Africa: Statistical analyses and sensitivity experiments. Climate Dynamics , 14, 353-368.

[28] Fouquart, Y, & Bonnel, B. (1980). Computations of solar heating of the earth's atmosphere: a new parameterization. Beitr Phys. Atmos. , 53, 35-62.

[29] Fritsch, J. M, & Chappell, C. F. (1980). Numerical prediction of convectively driven mesoscale pressure systems. Part I: Convective parameterization. Journal of the Atmospheric Sciences , 37, 1722-1733.

[30] Gaetani, M, Fontaine, B, Roucou, P, & Baldi, M. (2010). Influence of the Mediterranean sea on the West African monsoon: Intraseasonal variability in numerical simulations. Journal of Geophysical Research 115: D24115, doi:JD014436

[31] Gallee, H, Moufouma-okia, W, Bechtold, P, Brasseur, O, Dupays, I, Marbaix, P, Messager, C, Ramel, R, & Lebel, T. (2004). A high resolution simulation of a West African rainy season using a regional climate model. Journal of Geophysical Research, doi:JD004020.

[32] Garand, L. (1983). Some improvements and complements to the infrared emissivity algorithm including a parameterization of the absorption in the continuum region. Journal of the Atmospheric Sciences , 40, 230-244.

[33] Gaye, A. T, Viltard, A, & De Felice, P. (2005). Lignes de grains et pluies en Afrique de l'Ouest: part des lignes de grains `a la pluie totale des´et´es 1986 et 1987. Sécheresse , 16, 269-273.

[34] Giannini, A, Saravanan, R, & Chang, P. (2003). Oceanic forcing of Sahel rainfall on interannual to interdecadal time scales. Science doi:10.1126/science.1089357. , 302, 1027-1030.

[35] Giorgetta, M, & Wild, M. (1995). The water vapor continuum and its representation in ECHAM4, Rep. 162, 38 pp., Max-Planck-Inst. für Meteorol, Hamburg, Germany.

[36] Giorgi, F, & Mearns, L. O. (1999). Introduction to special section: regional climate modelling revisited. Journal of Geophysical Research , 104, 6335-6352.

[37] Giorgi, F, Jones, C, & Asrar, G. (2009). Addressing climate information needs at the regional level: the CORDEX framework. World Meteorology Organ Bulletin Available online at http://wcrp.ipsl.jussieu.fr/RCD_Projects/CORDEX/CORDEX_giorgi_WMO.pdf. , 58, 175-183.

[38] Giorgi, F, Coppola, E, Solmon, F, Mariotti, L, Sylla, M. B, Bi, X, Elguindi, N, Diro, G. T, Nair, V, Giuliani, G, Cozzini, S, Guettler, I, Brien, O, Tawfik, T, Shalaby, A, Zakey, A, Steiner, A. S, Stordal, A, Sloan, F, Brankovic, L, & Reg, C. CM4: Model description and preliminary tests over multiple CORDEX domains. Climate Research doi:cr01018., 52, 7-29.

[39] Grasselt, R, Schüttemeyer, D, Warrach, K. S, Ament, F, & Simmer, C. (2008). Validation of TERRA-ML with discharge measurements, Meteorol. Z., 17(6), 763-773, doi: 10.1127/

[40] Grell, G. A. (1993). Prognostic evaluation of assumptions used by cumulus parameterizations, Monthly Weather Review , 121, 764-787.

[41] Gregory, D, & Rowntree, P. R. convection scheme with representation of cloud ensemble characteristics and stability dependent closure. Journal of Geophysical Research , 92, 14-198.

[42] Gregory, D, & Allen, S. (1991). The effect of convective downdraughts upon NWP and climate simulations. Proc. Ninth conference on numerical weather prediction, Denver, Colorado, , 122-123.

[43] Hagos, S. M, & Cook, K. H. (2007). Dynamics of the West African monsoon jump. Journal of Climate , 20, 52-64.

[44] Hagemann, S. (2002). An improved land surface parameter dataset for global and regional climate models. MPI Report 336: 21 pp.

[45] Hernández-díaz, L, Laprise, R, Sushama, L, Martynov, A, Winger, K, & Dugas, B. (2012). Climate simulation over the CORDEX-Africa domain using the fifth generation Canadian Regional Climate Model (CRCM5). Climate Dynamics (in press), doi:s00382-012-1387-z.

[46] Hourdin, F, Mustat, I, Guichard, F, Ruti, P. M, & Favot, F. Pham MMA, Grandpeix JY, Polcher J, Marquet P, Boone A, Lafore JP, Redelsperger JL, Dell'Aquilla A, Losada Doval T, Traore AK, Gallee H. (2010). AMMA-Model intercomparison project. Bulletin of the American Meteorology Society, doi:BAMS2791.1.

[47] Hsie, E. Y, Anthes, R. A, & Keyser, D. (1984). Numerical simulation of frontogenesis in a moist atmosphere. Journal of the Atmospheric Sciences , 41, 2581-2594.

[48] Hsieh, J. S, & Cook, K. H. (2005). Generation of African easterly wave disturbances: relationship to the African easterly jet. Monthly Weather Review , 133, 1311-1327.

[49] Hsieh, J. S, & Cook, K. H. (2007). A study of the energetic of African easterly waves using regional climate model. Journal of the Atmospheric Sciences , 64, 421-440.

[50] Hsieh, J. S, & Cook, K. H. (2008). On the instability of the African easterly jet and the generation of African waves: reversals of the potential vorticity gradient. Journal of the Atmospheric Sciences , 65, 2130-2151.

[51] Hudson, D. A, & Jones, R. (2002). Regional climate model simulations of present-day and future climates of southern Africa. Technical Note 39. Hadley Centre for Climate Prediction and Research, Met Off Bracknell, England.

[52] Jacob, D, Andrae, U, Elgered, G, Fortelius, C, Graham, L. P, Jackson, S. D, & Karstens, U. Chr.Koepken, Lindau R, Podzun R, Rockel B, Rubel F, Sass HB, Smith RND, Van

den Hurk BJJM, Yang X. (2001). A Comprehensive Model Intercomparison Study Investigating the Water Budget during the BALTEX-PIDCAP Period. Meteorology and Atmospheric Physics 77 (1-4): 19- 43.

[53] Jenkins, G. S, Gaye, A. T, & Sylla, B. (2005). Late 20th Century attribution of drying trends in the Sahel from the Regional Climate Model (RegCM3). Geophysical Research Letters 32, L22705, doi:GL024225.

[54] Jones, C, Giorgi, F, & Asrar, G. (2011). The Coordinated Regional Downscaling Experiment: CORDEX An international downscaling link to CMIP5. CLIVAR Exchanges , 56(16), 34-41.

[55] Jones, R. G, Murphy, J. M, & Noguer, M. (1995). Simulation of climate change over Europe using a nested regional climate model I: assessment of control climate, including sensitivity to location of lateral boundary conditions. Quarterly Journal of the Royal Meteorology Society , 121, 1413-1449.

[56] Jones, R, Noguer, M, Hassel, D, Hudson, D, Wilson, S, Jenkins, G, & Mitchell, J. (2004). Generating high resolution regional climate change using PRECIS, Met Office Hadley Centre. Exeter, UK, , 40.

[57] Kain, J. S, & Fritsch, J. M. entraining/detraining plume model and its application in convective parameterization. Journal of the Atmospheric Sciences , 47, 2784-2802.

[58] Kalnay, E, Kanamitsu, M, Kistler, R, Collins, W, Deaven, D, Gandin, L, Iredell, M, Saha, S, White, G, Woollen, J, Zhu, Y, Leetmaa, A, Reynolds, R, Chelliah, M, Ebisuzaki, W, Higgins, W, Janowiak, J, Mo, K. C, Ropelewski, C, Wang, J, Jenne, R, & Joseph, D. (1996). The NCEP- NCAR 40- year reanalysis project. Bull Amer Soc , 77, 437-471.

[59] Kamga, A. F, & Buscarlet, E. (2006). Simulation du climat de l'Afrique de l'Ouest à l'aide d'un modèle climatique régional. La Météorologie , 52, 28-37.

[60] Kessler, E. (1969). On the distribution and continuity of water substance in atmospheric circulation models, Meteorol. Monogr., 10 (32). American Meteorology Society. Boss Mass

[61] Kiehl, J. T, Hack, J. J, Bonan, G. B, Boville, B. A, Briegleb, B. P, Williamson, D. L, & Rasch, P. J. (1996). Description of the NCAR Community Climate Model (CCM3). NCAR Tech. Note 4201STR, 152 pp.

[62] Kjellström, E, Bärring, L, Gollvik, S, & Co-authors, A. years simulation of European climate with the new version of the Rossby Centre regional atmospheric climate model (RCA3). Reports Meteorology and Climatology 108, SMHI, SE-60176, Norrköping, Sweden., 140.

[63] Konaré, A, Zakey, A. S, Solmon, F, Giorgi, F, Rauscher, S, Ibrah, S, & Bi, X. (2008). A regional climate modeling study of the effect of desert dust on the West African monsoon. Journal of Geophysical Research 113: D12206, doi:10.1029/2007JD009322.

[64] Le Barbe LLebel T, Tapsoba D. (2002). Rainfall variability in West Africa during the years 1950-1990. Journal of Climate , 15, 187-202.

[65] Legates, D. R, & Willmott, C. J. (1990). Mean seasonal and spatial variability in gauge-corrected, global precipitation. International Journal of Climatology, , 10, 111-127.

[66] Lenderink, G. van den Hurk B, van Meijgaard E, van Ulden A, Cuijpers J. (2003). Simulation of present-day climate in RACMO2: first results and model developments. KNMI Technical Report (252) 24 pp.

[67] Leroux, S. Hall NMJ. (2009). On the relationship between African easterly waves and the African easterly jet. Journal of the Atmospheric Sciences doi:JAS2988.1., 66, 2303-2316.

[68] Lin, Y. L, Farley, R. D, & Orville, H. D. (1983). Bulk parameterization of the snow field in a cloud model. Journal of Climatology and Applied Meteorology , 22, 1065-1095.

[69] Lohmann, U, & Roeckner, E. (1996). Design and performance of a new cloud microphysics scheme developed for the ECHAM general circulation model. Climate Dynamics , 12, 557-572.

[70] Lu, J, & Delworth, R. (2005). Oceanic forcing of the late 20th century Sahel drought. Geophysical Research Letter 32, L22706, doi:GL023316.

[71] Mariotti, L, Coppola, E, Sylla, M. B, Giorgi, F, & Piani, C. (2011). Regional climate model simulation of projected 21st century climate change over an all-Africa domain: Comparison analysis of nested and driving model results. Journal of Geophysical Research 116, D15111, doi:JD015068.

[72] Mathon, V, & Laurent, H. (2001). Life cycle of Sahelian mesoscale convective cloud systems. Quarterly Journal of the Royal Meteorology Society , 127, 377-406.

[73] Mcgregor, J. L, Katzfey, J. J, & Nguyen, K. C. (1998). Fine resolution simulations of climate change for Southeast Asia. Final report for a Research Project commissioned by Southeast Asian Regional Committee for START (SARCS), Aspendale, Vic.. CSIRO Atmospheric Research VI: , 15-35.

[74] Mekonnen, A, Thorncroft, C. D, & Aiyyer, A. (2006). Analysis of convection and its association with African easterly waves. Journal of Climate , 19, 5405-5421.

[75] Messager, C, Gallee, H, & Brasseur, O. (2004). Precipitation sensitivity to regional SST in a regional climate simulation during the West African monsoon for two dry years. Climate Dynamics , 22, 249-266.

[76] Mitchell, T. D, Carter, T. R, Jones, P. D, Hulme, M, & New, M. (2004). A Comprehensive Set of High-Resolution Grids of Monthly Climate for Europe and the Globe: The

Observed Record (1901-2000) and 16 Scenarios (2001-2100). Tyndall Centre for Climate Change Research, Norwich, U.K., Working Paper 55.

[77] Mohr, K. I, & Thorncroft, C. D. (2006). Intense convective systems in West Africa and their relationship to the African easterly jet. Quarterly Journal of the Royal Meteorological Society , 132, 163-176.

[78] Morcrette, J. J. (1991). Radiation and cloud radiative properties in the ECMWF forecasting system, Journal of Geophysical Research, 96(D5): 9121-9132, doi:JD01597.

[79] Morcrette, J. J, Smith, L, & Fouquart, Y. (1986). Pressure and temperature dependence of the absorption in longwave radiation parameterizations. Beitr Physics Atmos. , 59, 455-469.

[80] Moufouma-okia, W, & Rowell, D. P. (2010). Impact of soil moisture initialisation and lateral boundary conditions on regional climate model simulations of the West African Monsoon. Climate Dynamics doi:s00382-009-0638-0., 35, 213-229.

[81] Murthi, A, Bowman, K. P, & Leung, L. R. (2011). Simulations of precipitation using NRCM and comparisons with satellite observations and CAM: annual cycle. Climate Dynamics doi:s00382-010-0878-z., 36, 1659-1679.

[82] Nicholson, S. E. (2008). The intensity, location and structure of the tropical rainbelt over West Africa as factors in inter annual variability. International Journal of Climatology , 28, 1775-1785.

[83] Nicholson, S. E, & Grist, J. P. (2001). A conceptual model for understanding rainfall variability in the West African Sahel on interannual and interdecadal timescales. International Journal of Climatology , 21, 1733-1757.

[84] Nicholson, S. E, Barcilon, A. I, & Challa, M. (2008). An analysis of West African dynamics using a linearized GCM. Journal of Atmospheric Science doi: 10.1175/2007JAS2194.1. , 64, 1182-1203.

[85] Nikulin, G, Jones, C, Samuelsson, P, Giorgi, F, Asrar, G, Büchner, M, Cerezo-mota, R, Christensen, O. B, Déqué, M, Fernandez, J, Hänsler, A, Van Meijgaard, E, Sylla, M. B, & Sushama, L. (2012). Precipitation Climatology in an Ensemble of CORDEX-Africa Regional Climate Simulations. Journal of Climate doi:JCLI-D-11-00375.1, 6057-6078.

[86] Nordeng, T. E. (1994). Extended versions of the convective parametrization scheme at ECMWF and their impact on the mean and transient activity of the model in the tropics. Research Department.

[87] Paeth, H, Born, K, Podzun, R, & Jacob, D. (2005). Regional dynamical downscaling over West Africa: model evaluation and comparison of wet and dry years. Meteorol Z , 14(3), 349-367.

[88] Paeth, H, Born, K, Grimes, R, Podzun, R, & Jacob, D. (2009). Regional climate change in tropical and northern Africa due to greenhouse forcing and land use changes. Journal of Climate, doi:JCLI2390.1.

[89] Paeth, H. Hall NMJ, Gaertner MA, Alonso MD, Moumouni S, Polcher J, Ruti PM, Fink AH, Gosset M, Lebel T, Gaye AT, Rowell DP, Moufouma-Okia W, Jacob D, Rockel B, Giorgi F, Rummukainen M. (2011). Progress in regional downscaling of West African precipitation. Atmospheric Science Letter doi:asl.306., 12, 75-82.

[90] Pal, J. S, Giorgi, F, Bi, X, Elguindi, N, Solmon, F, Gao, X, Francisco, R, Zakey, A, Winter, J, Ashfaq, M, Syed, F, Bell, J. L, Diffanbaugh, N. S, Kamacharya, J, Konare, A, & Martinez, D. da Rocha RP, Sloan LC, Steiner A. (2007). The ICTP RegCM3 and RegCNET: regional climate modeling for the developing world. Bulletin of American Meteorological Society , 88, 1395-1409.

[91] Pal, J. S, & Small, E. E. Eltahir EAB. (2000). Simulation of regional-scale water and energy budgets: representation of subgrid cloud and precipitation processes within RegCM. Journal of Geophysical Research , 105, 29579-29594.

[92] Patricola, C. M, & Cook, K. H. (2008). Atmosphere/vegetation feedbacks: a mechanism for abrupt climate change over northern Africa. Journal of Geophysical Research 113, doi:JD009608.

[93] Patricola, C. M, & Cook, C. H. (2010). Northern African climate at the end of the 21 twenty-first century: an integrated application of regional and global climate models. Climate Dynamics doi:s00382-009-0623-7., 35, 193-212.

[94] Pohl, B, & Douville, H. (2011). Diagnosing GCM errors over West Africa using relaxation experiments. Part I: Summer monsoon climatology and interannual variability. Climate Dynamics, doi:s00382-010-0911-2., 37, 1293-1312.

[95] Ramel, R, Gallee, H, & Messager, C. (2006). On the northward shift of the West African monsoon. Climate Dynamics, doi:s00382-005-0093-5.

[96] Rasch, P. J, & Kristjánsson, J. E. (1998). A comparison of the CCM3 model climate using diagnosed and predicted condensate parameterizations. Journal of Climate , 11, 1587-1614.

[97] Rechid, D, Raddatz, T. J, & Jacob, D. (2009). Parameterization of snow-free land surface albedo as a function of vegetation phenology based on MODIS data and applied in climate modelling. Theoretical and Applied Climatology , 95, 245-255.

[98] Redelsperger, J. L, Diongue, A, Diedhiou, A, Ceron, J. P, Diop, M, Gueremy, J. F, & Lafore, J. P. (2002). Multi-scale description of a Sahelian synoptic weather system representative of the West African Monsoon. Quaternary Journal of Royal Meteorological Society , 128, 1229-1257.

[99]   Redelsperger, J. L, Thorncroft, C. D, Diedhiou, A, Lebel, T, Parker, D. J, & Polcher, J. (2006). African Monsoon Multidisciplinary analysis: an international research project and field campaign. Bulletin of American Meteorology Society , 87, 1739-1746.

[100] Ritter, B, & Geleyn, J. F. (1992). A comprehensive radiation scheme of numerical weather prediction with potential application to climate simulations. Monthly Weather Review , 120, 303-325.

[101] Rockel, B, Will, A, & Hense, A. (2008). The Regional Climate Model COSMO-CLM (CCLM). Meteorolog. Z., , 17, 347-248.

[102] Rummukainen, M. (2010). State-of-the-art with regional climate Models. Climate Change , 1, 82-86.

[103] Ruti, P. M. Dell'Aquila A. (2010). The twentieth century African easterly waves in re-analysis systems and IPCC simulations, from intra-seasonal to inter-annual variability. Climate Dynamics , 35, 1099-1117.

[104] Samuelsson, P, Gollvik, S, & Ullerstig, A. (2006). The land-surface scheme of the Rossby Centre regional atmospheric climate model (RCA3). SMHI Rep. Met. 122, 25pp.

[105] Sànchez, E, Gallardo, C, Gaertner, M, Arribas, A, & Castro, M. (2004). Future climate extreme events in the Mediterranean simulated by a regional climate models: A first approach. Global and Planetary Change doi:s00382-001-0218-4., 44, 163-180.

[106] Sass, B. H, Rontu, L, Savijärvi, H, & Räisänen, P. (1994). HIRLAM-2 Radiation scheme Documentation and tests. SMHI HIRLAM Tech. Rep. 16, 43 pp.

[107] Savijärvi, H. (1990). A fast radiation scheme for mesoscale model and short-range forecast models. Journal of Applied Meteorology , 29, 437-447.

[108] Semazzi HFMSun L. (1997). The role of orography in determining the Sahelian climate. International Journal of Climatology , 17, 581-596.

[109] Sijikumar, S, Roucou, P, & Fontaine, B. (2006). Monsoon onset over Sudan-Sahel: Simulation by the regional scale model MM5. Geophysical Research Letter 33: L03814, doi:GL024819.

[110] Skinner, B. C, Ashfaq, M, & Diffenbaugh, N. S. (2012). Influence of twenty-first-century atmospheric and sea surface temperature forcing on West African climate. Journal of Climate (In Press), doi:JCLI4183.1.

[111] Smith RNB (1990). A scheme for predicting layer clouds and their water content in a general circulation model. Quaterly Journal of the Royal Meteorology Society. , 116, 435-460.

[112] Solmon, F, Elguindi, N, & Mallet, M. (2012). Radiative and climatic effects of dust over West Africa, as simulated by a regional climate model. Climate Research. doi:cr01039., 52, 97-113.

[113] Steiner, A. L, Pal, J. S, Rauscher, S. A, Bell, J. L, Diffenbaugh, N. S, Boone, A, Sloan, L. C, & Giorgi, F. (2009). Land surface coupling in regional climate simulations of the West African monsoon. Climate Dynamics doi:s00382-009-0543-6., 33, 869-892.

[114] Sultan, B, & Janicot, S. (2003). The West African monsoon dynamics. Part II: The "preonset" and "onset" of the summer monsoon. Journal of Climate , 16, 3407-3427.

[115] Sundquist, H. (1978). A parameterization scheme for non-convective condensation including precipitation including prediction of cloud water content. Quarterly Journal of the Royal Meteorology Society. , 104, 677-690.

[116] Sylla, M. B, Gaye, A. T, Pal, J. S, Jenkins, G. S, & Bi, X. Q. (2009). High resolution simulations of West Africa climate using Regional Climate Model (RegCM3) with different lateral boundary conditions. Theoretical and Applied Climatology 98 (3-4): 293-314, doi:s00704-009-0110-4.

[117] Sylla, M. B, Gaye, A. T, Jenkins, G. S, Pal, J. S, & Giorgi, F. (2010a). Consistency of projected drought over the Sahel with changes in the monsoon circulation and extremes in a regional climate model projections. Journal of Geophysical Research 115, D16108, doi:JD012983.

[118] Sylla, M. B, Coppola, E, Mariotti, L, Giorgi, F, & Ruti, P. M. Dell'Aquila A, Bi X. (2010b). Multiyear simulation of the African climate using a regional climate model (RegCM3) with the high resolution ERA-interim reanalysis. Climate Dynamics doi:s00382-009-0613-9., 35, 231-247.

[119] Sylla, M. B. Dell'Aquila A, Ruti PM, Giorgi F. (2010c). Simulation of the Intraseasonal and the Interannual Variability of Rainfall over West Africa with a Regional Climate Model (RegCM3) during the Monsoon Period. International Journal of Climatology doi:joc.2029., 30, 1865-1883.

[120] Sylla, M. B, Giorgi, F, Ruti, P. M, & Calmanti, S. Dell'Aquila A. (2011). The impact of deep convection on the West African summer monsoon climate: a regional climate model sensitivity study. Quarterly Journal of Royal Meteorological Society doi:qj. 853., 137, 1417-1430.

[121] Sylla, M. B, Gaye, A. T, & Jenkins, G. S. (2012a). On the Fine-Scale Topography Regulating Changes in Atmospheric Hydrological Cycle and Extreme Rainfall over West Africa in a Regional Climate Model Projections. International Journal of Geophysics 2012, Article ID 981649, doi:

[122] Sylla, M. B, Giorgi, F, & Stordal, F. origins of rainfall and temperature bias in high resolution simulations over Southern Africa. Climate Research doi:cr01044., 52, 193-211.

[123] Sylla, M. B, Giorgi, F, Coppola, E, & Mariotti, L. (2012c). Uncertainties in daily rainfall over Africa: assessment of observation products and evaluation of a regional climate model simulation. International Journal of Climatology. In Press, doi:joc.3551.

[124] Thorncroft, C. D, & Blackburn, M. (1999). Maintenance of the African easterly jet. Quaternary Journal of Royal Meteorological Society , 125, 763-786.

[125] Thorhcroft, C. D. Hall NMJ, Kiladis GN. (2008). Three-dimensional structure and dynamics of African Easterly Waves. Part III: Genesis. Journal of the Atmospheric Sciences doi:JAS2575.1., 65, 3596-3607.

[126] Thorncroft, C. D, Nguyen, H, Zhang, C, & Peyrillé, P. (2011). Annual cycle of the West African monsoon: regional circulations and associated water vapour transport. Quarterly Journal of the Royal Meteorology Society doi:qj.728., 137, 129-147.

[127] Tiedtke, M. (1989). A comprehensive mass flux scheme for cumulus parameterization in large scale models, Monthly Weather Review. , 117, 1779-1800.

[128] Tiedtke, M. (1993). Representation of clouds in large-scale models. Monthly Weather Review. , 121, 3040-3061.

[129] Van Den Hurk BJJMVan Meijgaard E. (2010). Diagnosing land-atmosphere interaction from a regional climate model simulation over West Africa. Journal of Climate, doi:JHM1173.1

[130] Van den Hurk BJJMViterbo P, Beljaars ACM, Betts AK. (2000). Offline validation of the ERA40 surface scheme, ECMWF Tech. Memo. 295, Eur. Cent. for Medium-Range Weather Forecasts, Reading, U. K.

[131] Vanvyve, E, Hall, N, Messager, C, Leroux, S, & Van Ypersele, J. P. (2008). Internal variability in a regional climate model over West Africa. Climate Dynamics doi:s00382-007-0281-6., 30, 191-202.

[132] Vizy, E, & Cook, K. (2002). Development and application of a mesoscale climate model for the tropics: influence of sea surface temperature anomalies on the West African monsoon. Journal of Geophysical Research 107(D3):4023, doi:JD000686

[133] Wang, G, & Alo, C. A. (2012). Change in precipitation seasonality in West Africa predicted by RegCM3 and the impact of dynamic vegetation feedback. International Journal of Geophysics. Article ID 597205, 10 pages, doi:

[134] Wang, G. Eltahir EAB. (2000). Ecosystem dynamics and the Sahel drought. Geophysical Research Letters , 27(6), 795-798.

[135] Xue, Y, & De Sales, F. Lau KMW, Bonne A, Feng J, Dirmeyer P, Guo Z, Kim KM, Kitoh A, Kumar V, Poccard-Leclercq I, Mahowald N, Moufouma-Okia W, Pegion P, Rowell DP, Schemm J, Schulbert S, Sealy A, Thiaw WM, Vintzileos A, Williams SF, Wu ML. (2010). Intercomparison of West African Monsoon and its variability in the West African Monsoon Modelling Evaluation Project (WAMME) first model Intercomparison experiment. Climate Dynamics dois00382-010-0778-2., 35, 3-27.

[136] Zaroug MAHSylla MB, Giorgi F, Eltahir EAB, Aggarwal PK. (2012). A sensitivity study on the role of the Swamps of Southern Sudan in the summer climate of North Africa using a regional climate model. Theoretical and Applied Climatology In press, doi:s00704-012-0751-6.

Climate Risk Assessment for Water Resources Development in the Niger River Basin Part II: Runoff Elasticity and Probabilistic Analysis

J.G. Grijsen, A. Tarhule, C. Brown, Y.B. Ghile, Ü. Taner, A. Talbi-Jordan, H. N. Doffou, A. Guero, R. Y. Dessouassi, S. Kone, B. Coulibaly and N. Harshadeep

Additional information is available at the end of the chapter

1. Introduction

The preceding chapter presented the context and overview of the climate risk assessment (CRA) for the Niger River Basin (NRB). It also discussed the climate change estimation methodology and the impact of potential runoff changes on the performance indicators of the Niger Basin Sustainable Development Action Plan (SDAP). In this chapter, we describe the methodology used in estimating the runoff response to climate change, followed by the quantitative assessment of climate risks for key water related sectors, and a discussion of the major findings.

2. Runoff response to climate change

For this study, the primary goal is to determine the relative changes (in %) in annual runoff and various performance criteria such as hydro-energy and irrigated agriculture production, due to relative changes (in %) in annual climate parameters (notably the precipitation, P, and temperature, T) caused by projected future climate changes. The task, then, is to determine the response of runoff to changes in climate parameters, i.e. to estimate the climate elasticity of runoff. A precipitation elasticity of runoff of, for example, 2.5 implies that a 10% increase in precipitation causes a 25% increase in runoff; a temperature elasticity of -0.5 implies that a 10% increase in temperature causes a 5% decrease in runoff. For this purpose we have extensively reviewed available literature on climate elasticity of basin runoff, *inter alia* Wigley and Jones (1985), Gleick (1986, 1987), Karl and Riebsame (1989), Risbey and Entekhabi (1996), Vogel et al (1999), Arora (2002), Legates et al (2005), including the estimation of climate change impacts on runoff through hydrological modeling studies for similar river basins in Africa, *inter alia* Deksyos Tarekegn (2006), Strzepek and McCluskey (2006) and SNC Lavalin (2007). We also applied linear and log-linear regression models to the available rainfall, temperature and runoff data for various sub-catchments of the Niger Basin (Grijsen and Brown, 2013). The data used were spatially aggregated (gridded) annual precipitation and temperature data for the period 1948-2002 for major sub-catchments of the Niger Basin (Hirabayashi, 2008) and annual stream flow data from the data base of the Niger Basin Observatory of the Niger Basin Authority (NBA). The Basin's network of hydrometric stations is shown in Fig. 1. In the preceding chapter, we showed that the projected changes in annual precipitation and temperature were well represented by the normal distribution.

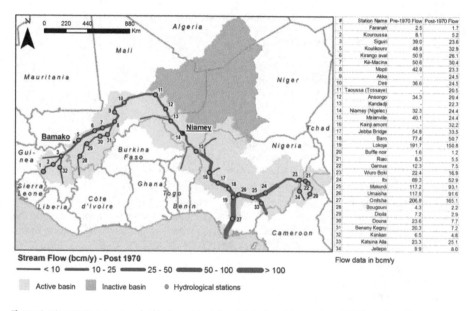

Figure 1. Niger Basin river network of hydrometric stations (Note: flow data are shown in billion m³/year)

Arora (2002) used the aridity index, $\varphi = E_0/P$, i.e. the ratio of annual potential evapotranspiration (E_0) to precipitation (P), to assess climate change impacts on annual runoff. Simple analytic expressions based solely on the aridity index of a basin are used to estimate changes in runoff due to changes in precipitation and (temperature driven) changes in potential evapotranspiration, as a first order estimate of the effect of climate change on annual runoff. The aridity index, φ, has been shown to describe the actual evaporation ratio E/P and the runoff ratio Q/P (= 1-E/P) of catchments for a range of climatic regimes. Arora (2002) used the aridity index to obtain analytic equations, which can be used to estimate relative changes in annual runoff due to relative changes in annual precipitation and available energy, i.e. the precipitation elasticity ε_P (= $[dQ/Q]/[dP/P]$ and the evaporation elasticity ε_{E0} (= $[dQ/Q]/[dE_0/E_0]$) of runoff. The latter is used to derive the temperature elasticity of runoff, ε_T = $[dQ/Q]/[dT/T]$. Because of its mathematical convenience, we used the functional form introduced by Turc (1954) and Pike (1964), as follows:

$E/P = [1+ \varphi^{-2}]^{-0.5}$ (actual evaporation ratio)

$Q/P = 1 - E/P$ (runoff coefficient)

$dQ/Q = (1 + \beta) \, dP/P - \beta \, dE_0/E_0$

$\varepsilon_P = [dQ/Q] \, / \, [dP/P] = 1 + \beta$ (precipitation elasticity of runoff)

$\varepsilon_{E0} = [dQ/Q] \, / \, [dE_0/E_0] = -\beta$ (potential evapotranspiration elasticity of runoff)

$\varepsilon_T = [dQ/Q]/[dT/T] = \varepsilon_{E0} \, T/(T+17.8) = -0.60\beta$ (temp. elasticity of runoff; T=26.6^{0}C; Hargreaves, 1982, 1985)

$\beta = [1 + \varphi^2]^{-1} \, / \, \{[1 + \varphi^{-2}]^{0.5} - 1\} = (1 + E/P) \, E/P = 2 - 3 \, Q/P + (Q/P)^2$

Thus, the (observed) runoff coefficient provides a simple initial estimator for the climate elasticities of runoff for the Niger Basin (T=26.6^{0}C), i.e.:

$\varepsilon_P = 3 - 3 \, Q/P + (Q/P)^2$

$\varepsilon_T = -1.2 + 1.8 \, Q/P - 0.6 \, (Q/P)^2$

2.1. Results of runoff elasticity analysis

We applied linear regression analysis on the historical relative variations (in %) in annual precipitation and runoff for multiple sub-catchments of the Niger Basin, and compared actual and theoretical values of the aridity index φ and precipitation elasticity ε_P based on the runoff coefficient Q/P (Fig. 2). Overall the runoff regime of the Niger Basin (i.e. for the runoff generating sub-catchment in the Upper Niger and Benue Basins) is well represented by a runoff coefficient Q/P = 0.2, which corresponds to a precipitation elasticity of runoff ε_P = 2.44 and temperature elasticity ε_T = -0.86. The former value agrees well with the results of our regression analyses, and we have thus adopted the precipitation elasticity ε_P = +2.5 for our study.

Hydrological modeling studies for similar basins in Africa generally showed lower values for the temperature elasticity of runoff, and we have thus adopted ε_T = -0.75[1]. Thus, a 10% (2.7^{0}C) increase in temperature (causing a 7.5% decline in runoff) and a concomitant 3% increase in precipitation (causing a 7.5% increase in runoff) would yield no net change in runoff. Projections of future precipitation and temperature were subsequently translated into annual basin runoff for the ensemble of 38 climate projections for the 21st century, centered on 2030, 2050 and 2070 (Fig. 2), by using the precipitation - temperature - runoff regression model of the form:

---

1 The temperature elasticity of -0.75 corresponds to a temperature sensitivity of run equal to -0.75/T, or approximately -3% runoff per 0C temperature increase.

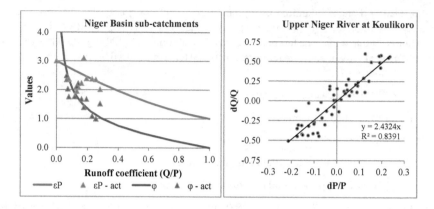

Figure 2. Theoretical and actual ε_P and φ parameter values for NRB sub-catchments, based on the runoff coefficient (left panel; blue and red lines represent theoretical values) and linear regression of historical changes in annual precipitation and runoff for the Upper Niger River at Koulikoro (right panel)

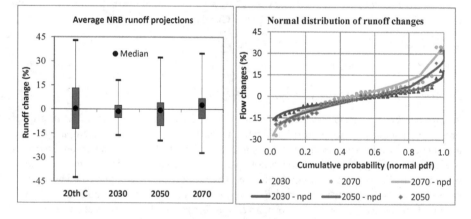

Figure 3. Runoff quartiles and probability distributions for projected runoff changes (2030, 2050 and 2070)

$$dQ/Q_0 = \varepsilon_P \, P/P_0 + \varepsilon_T \, dT/T_0$$

The projected mean flows are essentially constant over the 21st century at an average of 2% below the 20th century mean till 2050, with an insignificant increase of 1% by 2070; the standard deviation of projected mean flows is 7% by 2030 and 13% by 2070, reflecting the increasing uncertainty in climate projections for the distant future. The probability of a decrease in average annual runoff is just over 50%. The probability of a 20% decrease in average runoff by 2050 is minimal, which could be considered as a worst case scenario for standard project economic analyses for SDAP, with 2050 as an investment horizon. The projected mean flows adhere well to the normal probability distribution (Fig. 3, right panel).

By 2050 the projected temperature increase (8% or 2.1°C) causes an average decrease of 6% in runoff due to increased evapotranspiration, which is partially compensated by a projected 4% additional runoff due to increased rainfall, yielding a net decrease in mean runoff of only about 2%. Since annual rainfall, temperature and runoff adhere to the normal distribution, we can estimate the distribution of projected changes in runoff from the expected average changes in rainfall and temperature, and the variances of these changes (as for example available from the Climate Portal and Climate Wizard), as follows:

$E\{dQ/\mu_Q\} = \varepsilon_P\, E\{dP/\mu_P\} + \varepsilon_T\, E\{dT/\mu_T\}$; $E\{...\}$ denotes the expected/projected shift in means of P and T

$CV^2(dQ/\mu_Q) = \varepsilon_P^2\, CV^2(dP/\mu_P) + \varepsilon_T^2\, CV^2(dT/\mu_T)$; $CV(...)$ denotes coefficients of variation of projections

The contribution of temperature variations to the variance of runoff is negligible due to its small CV compared to the CV of changes in precipitation. This further explains that the temperature elasticity cannot be estimated through regression analysis. Probability distributions of future annual runoff can thus be based on an estimation of the average change in runoff - based on projected changes in annual precipitation and temperature and climate elasticities of runoff - and a commensurate shift of the historical distribution of annual runoff, as long as the inter-annual variability of precipitation does not change.

While all analyses pointed to a precipitation elasticity of about +2.5, the choice of the temperature elasticity or sensitivity is less certain, yet critical for the outcome of climate risk assessment. Therefore, a sensitivity analysis was done for a precipitation elasticity of +2.5 and a temperature elasticity of -1.25^2 (instead of -0.75). Results indicate that by 2050 the increase of 8% in temperature (2.1°C) would cause indeed an additional decrease of runoff by 4% due to the difference in temperature elasticities.

3. Quantitative climate risks for key water related sectors

Climate risks to key sectors are estimated on the basis of projected future runoff changes, where climate risk is defined based on probabilities of selected percentage changes in performance metrics relative to baseline operations. Recall that the baseline development scenario was defined in the preceding chapter as the scenario in which the Fomi (FO), Toussa (TA) and Kandaji (KD) dams and associated irrigation infrastructure are fully implemented. As previously discussed, the relationships between relative changes in runoff and relative changes in selected performance criteria at basin level were derived from numerous Mike-Basin runs for variations in runoff between -30% and + 10% of the 20th century baseline conditions, and water demands with a 5% increase compared to the 20th century baseline (Table 3, preceding chapter). The runoff changes required to cause specific changes in selected performance indicators are shown in Fig. 11 (preceding chapter). The results shown in Fig.

---

2 The temperature elasticity of -1.25 corresponds to a temperature sensitivity of runoff equal to -1.25/T, or approximately 5% decrease in runoff per 0C increase in temperature, which is well above the values mostly found in literature.

11 for minimum flows are less realistic than the other results, since minimum flows are impacted mainly by changes in dry season flows and irrigation abstractions, rather than by changes in annual flows. They are indicative nonetheless of the sensitivity of minimum flows to climate changes.

Figure 4 shows probabilities of risks for the average (A) and 1/5 years (20% dry) perform-ance of selected indicators (risk levels defined as shown in Fig. 3 of the preceding chap-ter), estimated as follows. The relationships between specific changes in the performance indicators and changes in the runoff required to generate those specific changes in most indicators is nearly linear in the most relevant domain of -20% to 0% of change in runoff (Fig. 11 of the preceding chapter). Thus, the performance indicators will also be distribut-ed according to the normal probability distribution, similar to the projected changes in runoff. When we find e.g. a 25% probability that future mean flows (averaged over many years) will at least be 10% less than the present long-term mean flows, we also assume there is 25% probability that a specific performance indicator is at least xy% less than the historic indicator values; the xy% value is derived from the column '-10%R' in the performance matrix of Table 3 (preceding chapter). In reverse, to estimate the probability that e.g. projected hydro-energy generation will at least be 20% less than at present, we determine the runoff reduction required to cause 20% less hydro-energy (16% per Fig. 11 of the preceding chapter), and estimate the probability of such runoff reduction based on the normal distribution for the projected future runoff (2% in 2030 and 10% in 2070).

Figure 4 shows specific percentiles (5, 25, 50, 75 and 95%) of the selected performance indica-tors. The concerned percentiles of runoff changes were first determined from the normal probability distribution of projected future runoff changes; then the corresponding percentiles of each performance indicator were interpolated from the performance matrix in Table 3 of the preceding chapter.

The probability that by 2050 or 2070 one may see decreases of more than 20% (significant risk) in the various performance criteria is only 5 to 15% or less, other than for the minimum environmental flow conditions in the Inner Delta and Middle Niger, which are under severe risk. The probability that one may not be able to maintain the required minimum flows under the fully developed FO-TA-KD scenario is 100%. This will be primarily caused by (projected) increased water abstractions for irrigation at the large Office du Niger irriga-tion scheme in Mali. Hydro-energy, navigation and flooding of the Inner Delta vary similar to runoff variations, with runoff elasticities of about 1 to 1.2. There is 25 - 35% probabili-ty that by 2050 these sectors could suffer performance decreases above 10% (moderate risk). The probability that the performance decreases in these sectors would be between 10 and 20% is about 20% and the probability of performance decreases exceeding 20% is be-tween 5 and 15%. Impacts on irrigated agriculture are minimal under the prevailing priority of water allocation to agriculture. Figures 3 and 4 serve to illustrate the main findings and conclusions of this CRA, as highlighted in the following.

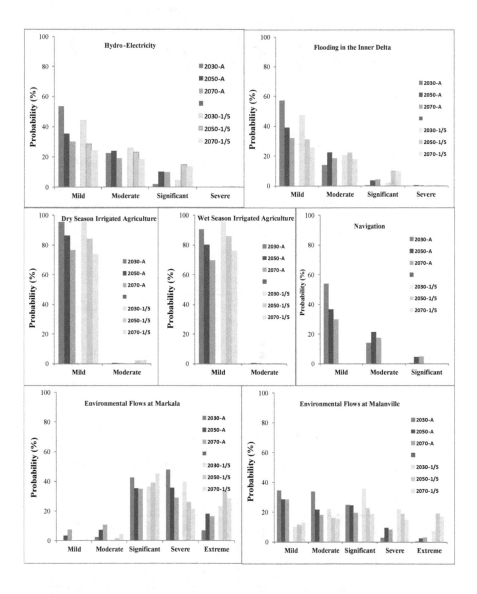

Figure 4. Probabilities of risks for average (A) and 1/5 years (20% dry) performance of selected indicators; risk levels are defined in Fig. 3 of the preceding chapter.

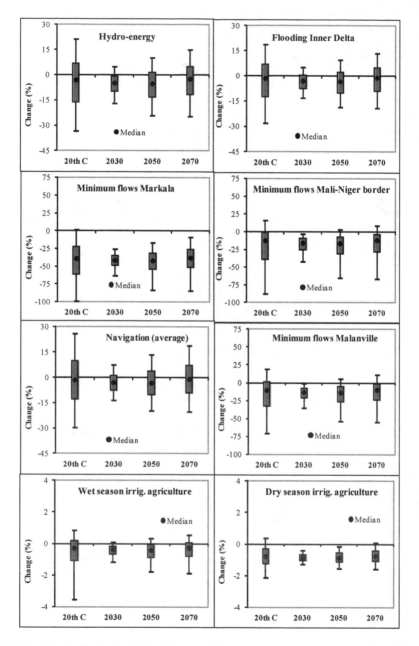

Figure 5. Percentiles (5, 25, 50, 75 and 95%) of selected performance indicators for an average year (Note: Minimum flow data refer to 10-day average flows)

3.1. Sensitivity of selected sectors to climate change

Irrigated agriculture is insensitive to projected climate changes; SDAP and particularly the construction of Fomi dam is an effective adaptation measure. Under SDAP total agricultural output in the NRB is projected to increase by nearly 450%. The current water allocation rules prioritize irrigation water demands in order to secure food production and alleviate poverty in the long term. This makes irrigated agriculture in the NRB insensitive to decreased runoff as a result of the projected climate changes, but it places pressures on the existing reservoir systems to provide a reliable water supply for hydro-electricity production, navigation, and environmental flows. These sectors are found to be more vulnerable to the risk of decreased runoff. Under the current water allocation rules, mild agricultural production decreases are likely to occur, but would generally be less than a few percentage points. As long as reservoirs can be fully replenished during the rainy season, climate change has little impact on the supply side of irrigation during the dry season. Indeed, by regulating the variability of the Upper Niger flow, SDAP and particularly the construction of Fomi dam will be good insurance for the protection of irrigated agriculture against the potential negative impacts of climate change.

Climate change impacts on hydro-energy, navigation, and Inner Delta Flooding are projected to be mild to moderate (less than 20% decrease). Hydro-energy, navigation and flooding of the Inner Delta vary similar to runoff variations, with runoff elasticities of about 1 to 1.2^2. There is 25 to 35% probability that by 2050 these sectors could suffer performance decreases more than 10%, and the probability of performance decreases exceeding 20% is only between 5 and 15%. Hence, risks of climate change impacts on these sectors will be mild to moderate. These impacts are similar to the declines in navigation and Inner Delta flooding caused by the full implementation of the FO-TA-KD development scenario (10-20%).

Climate change impacts on minimum flows can be severe. Minimum flows entering the Inner Delta (downstream of the Office du Niger irrigation scheme in Mali) are sensitive to an increase in water demands by for example 5% at the Office du Niger (ON) irrigation scheme, due to increased future temperatures and evapotranspiration. The design of SDAP has planned irrigation development and abstractions during dry season irrigation at the ON such that minimum flows at Markala dam can just be maintained at the present ON water demands per hectare for dry season irrigation. The probability that by 2050 - under the FO-TA-KD development scenario - one may not be able to maintain the required minimum flows through the Inner Delta and the Middle Niger to Malanville at the Niger-Nigeria border is 100%.

Minimum environmental flows during the dry season are the most sensitive to climate change (particularly with increased irrigation water demands) under the present water allocation rules, with runoff elasticities on the order of 2.5 to 4.0^3. By 2050 there is a 50% probability that once in 5 years the 10-day average minimum inflow to the Inner Delta will be less than 25 m³/s; i.e. less than 60% of the adopted norm of 40 m³/s. Relative decreases of minimum flows at the Mali-Niger border, Niamey and Malanville are slightly less severe.

---

3 The large runoff elasticities of minimum flows indicate that moderate decreases in annual runoff can cause severe reductions in the minimum flows at Markala and further downstream on the Middle Niger.

Adaptations required for enhancing minimum Niger River flows. Minimum flows are severely impacted by increased water demands of the Office du Niger (ON) irrigation scheme due to increased evapotranspiration. Should higher priority be accorded to the sustenance of minimum flows, as per the Water Charter agreed between the NBA member countries - dry season irrigated agriculture could become moderately sensitive to climate change impacts, since the abstractions for irrigation at ON – as planned under the FO-TA-KD scenario - would need to be reduced in favor of releasing more water into the Inner Delta. This potential problem can be addressed by measures to increase the present low dry season irrigation efficiency at ON (about 30%), and/or by slightly reducing the future dry season irrigated areas, such that minimum flows can also be maintained in the future. Future abstractions during the critical low flow period February to May would by 2050 amount on average to 425 m^3/s, and would need to be reduced by about 25 m^3/s (6%) to avoid unacceptable minimum environmental flows passing Markala dam. This can be achieved by implementing irrigation system rehabilitations and water management measures aimed at increasing the dry season irrigation efficiency at ON by about 2%; presently the dry season irrigation efficiency at ON is less than 30%. Other options to reduce abstractions by at least 6% during the dry season would be to reduce the acreage projected to be cropped in the dry season by 6%, or to reduce dry season rice planting in favor of an extension of less water demanding horticulture and other non-rice crops.

Worst case scenario for project economic analysis. The probability of a 20% decrease in average runoff in the Niger Basin is minimal, which could thus be considered a worst case scenario for the standard economic analysis of infrastructure projects under SDAP. For example, a detailed economic analysis of the Kandadji hydro-power and irrigation project - located on the Niger River in Niger near the Malian border - was performed as part of project appraisal (World Bank, 2012), yielding an Economic Internal Rate of Return (EIRR) of 13.5% for the base case (2005 hydrology). Sensitivity analyses showed that the EIRR would reduce to 12.4% for an overall decline of 20% in the runoff of the upstream basin, still above the World Bank's financing threshold rate of 12%. The runoff elasticity of EIRR was assessed at about 0.4, i.e. a 20% decrease in runoff causes an 8% decrease in the EIRR (1.1 percentage points down from 13.5%). Thus a 30% decline in average runoff would push the EIRR below the threshold rate of 12%.

3.2. Economic analysis of SDAP under climate change conditions

Potential climate change impacts on the economic performance of SDAP are modest. The economic analysis of climate change impacts on the full SDAP investment plan has focused on hydro-power development with associated irrigation development as the main drivers for SDAP. Various proposed Run-of-River (R-o-R) hydropower schemes were also included. Primary economic benefits from the FO-TA-KD Program and R-o-R schemes will accrue from the combination of hydropower generation and new schemes brought under irrigation. Secondary benefits include provision of water supply (particularly for towns and villages along the Middle Niger River), environmental benefits of wetlands (reservoir surface areas and newly irrigated rice planted areas), fish production in the new reservoirs, ecosystem regeneration in the Niger Valley, and livestock production. Economic and environmental losses will occur in the Inner Delta (rice production, environmental benefits of wetlands, fisheries and livestock), related to navigation in the Middle and Lower Niger, and due to reduced hydro-energy generation at Kainji and Jebba dams in Nigeria caused by upstream water diversions for irrigation and evaporation losses from the new reservoirs. Losses due to climate changes, which would also occur under the prevailing (2005) "natural conditions", such as losses of Inner delta flooded areas and losses of hydro-energy at Kainji and Jebba dams due to climate change, have been excluded from the analysis of climate change impacts on the SDAP Program. Results of the economic analysis for the SDAP, in terms of EIRR, are shown in Table 1, for the baseline (2005) hydrological conditions and for the situation with a 20% flow reduction due to climate change during the entire life of the SDAP project components (2015 – 2050).

Development scenario	EIRR (2005 hydrology)	EIRR under 20% runoff reduction	Runoff elasticity of EIRR
FO-TA-KD + R-o-R	13.0	11.3	0.64
FO-KD + R-o-R	14.6	13.0	0.55
FO-TA-KD; no R-o-R	9.5	7.9	0.84
FO-KD; no R-o-R	11.7	10.2	0.64
FO-TA-KD + R-o-R; no irrigation development	14.1	12.3	0.64
FO-TA-KD; no R-o-R; no irrigation development	10.1	8.0	1.0
KD + 45,000 ha irrigation (no R-o-R)	13.5	12.4	0.42

Table 1. Estimated EIRR for various SDAP development scenarios, with/without impacts of climate change

Potential climate change impacts on the economic performance of the SDAP Program are assessed as mild (<10%) to moderate (<20%). Under the above 'worst case' scenario of 20% reduction in the long-term average basin runoff, the projected climate change impacts on irrigated agriculture, navigation, minimum flows and flooding of the Inner Delta have only a minor impact on the overall economic performance of the SDAP. In economic terms the only significant impact of reduced runoff due to climate changes would stem from its direct impact on hydro-energy production. In the 'worst case scenario', with the 20% flow reduction imposed as from 2015, one can expect a 1.7% reduction in the EIRR of 13% for the SDAP under the base case (2005) hydrological conditions. The runoff elasticity of the EIRR for the SDAP is about 0.6, compared to 0.4 for the Kandadji (stand-alone) Program. The investment in Run-of-River hydropower schemes appears to be the most beneficial component of the SDAP, since it increases the EIRR of the FO-TA-KD scheme by 3.5%. It allows the SDAP to maintain a robust EIRR above 12%, and even more than 11% in the 'worst case' scenario. Without R-o-R schemes the EIRR drops well below 10% under conditions of 20% reduction in average runoff due to climate change.

Irrigated agriculture achieves an adequate EIRR as long as the cost of the new dams are treated as sunk cost, i.e. are fully charged to hydropower development. Losses to dry season irrigated agriculture under climate change, estimated at 8% for an average runoff reduction of 20%, cause only a loss of 0.1% of the EIRR under the 'worst case' runoff scenario. Similarly, the impacts of losses due to reduced flooding of the Inner Delta (including losses to wetlands, fisheries, floating rice production and livestock) caused by Fomi dam and the planned large extensions of the Office du Niger irrigation schemes have no significant impact on the EIRR, causing only a loss of 0.2% in the EIRR. Impacts of climate change on these losses are similar under the present (2005) as well as future (FO-TA-KD) hydrological conditions, and were thus excluded from the SDAP economic analysis. Finally, losses to navigation under the FO-TA-KD scenario cause only a reduction of 0.2% of the EIRR.

4. Conclusions

While the risk of climate change has generated many policy and adaptation initiatives aimed at combating it, few of these initiatives are based on quantitative assessments of expected impacts to specific sectors. We argue that such assessments are necessary to convince decision makers and as a means of prioritizing and targeting scarce resources at the most vulnerable sectors or those sectors likely to create the largest domino effect. This study represents an important contribution towards that goal. Beginning with a bottom-up definition of risk, we estimated the expected magnitude of climate change, principally precipitation and temperature changes, translated those changes into runoff changes, and then estimated the impacts of projected runoff changes on key development sectors in the Niger Basin with respect to thresholds of changes in specific performance indicators.

The results show that climate change impacts in the Niger Basin will be generally mild (<10%) to moderate (<20%). By 2050, average annual temperatures in the Niger Basin would rise by 2.1°C, an 8% increase over baseline values. The increase in temperature will increase evapotranspiration, leading to an average decrease of 6% in runoff. However, a projected 4% additional runoff due to increased rainfall will partially offset the decreased runoff, producing a net decrease in mean runoff of only about 2%. It is notable that the magnitude of these changes pale in comparison to the historical patterns of variability in precipitation and runoff, observed over the NRB in the 20th century. Even so, the results should be interpreted cautiously. In general, while climate projections do a good job on simulating the mean, they are generally incapable of reproducing variability around that mean. Thus, whereas the change in the mean may be small, variability around the mean may trigger shocks to economic and agro-ecological systems disproportionate to the change in mean conditions.

With specific respect to the Niger Basin SDAP and under the current water allocation rules, irrigated rainy season agriculture ('hivernage') remains insensitive to climate change even under severe reductions of runoff (e.g. up to 30%). Dry season irrigated agriculture ('contre-saison') and the sustenance of environmental flows during the low flow season are interlinked and respectively mildly and extremely vulnerable to severe reductions of runoff. However, with (i) the implementation of Fomi, Taoussa and Kandadji dams, along with (ii) optimal basin-wide reservoir management, (iii) increased irrigation efficiencies, particularly at the Office du Niger, (iv) gradual shifts from dry season rice to less water demanding non-rice dry season crops, and (v) other similar adaptation measures to improve water use efficiencies in particularly dry season irrigated agriculture, these 'sectors' can be well protected from the impacts of significant climate change.

Severe reductions in runoff would cause equally severe reductions in generated hydro-energy, navigation and flooding of the Inner Delta (projected to be mild to moderate under the present climate change projections). Such severe future impacts on (particularly) hydro-energy generation can only be minimized by reducing rainy season irrigated agriculture in the basin and/or by the construction of additional storage reservoirs along with hydro-energy generation facilities in the Upper Niger Basin and in Nigeria, particularly in the Benue Basin, which has mostly untapped hydro-power potential.

It is important to emphasize that in this report the risks are calculated based on the long term (i.e. 30 years) shift in mean precipitation and temperature values; they do not account for inter-annual or decadal variability. We do not consider this a serious limitation since the baseline period (1966-1988) - used also for the design of the FO-TA-KD scenario and simulated with the Mike Basin model - comprises arguably the most variable period in the historical data. It is noteworthy that the NRB has historically experienced runoff shortages greater than the projected levels. Thus, the historical experience provides an analogue for dealing with future climate; for water managers and farmers who do not know what to expect in the upcoming rainy season, managing the impacts of intra-seasonal, inter-annual variability of climate would be the priority to start with. Managing the near-term climate variability has also the potential to better prepare for dealing with long term-climate change impacts. Therefore, the use of seasonal to inter-annual hydrologic forecasts in reservoir operations and planning could be an important adaptation opportunity.

Acknowledgements

This study was done as part of the Climate Risk Assessment (CRA) for the Niger Basin, a joint initiative of the Niger Basin Authority (NBA) and the World Bank to assess the risks from climate change to the performance of NBA's Sustainable Development Action Plan (SDAP) for the Niger Basin. The aim of this initiative is to build resilience to climate risks into the SDAP. The Niger CRA study is supported by grants from (i) the Bank-Netherlands Partnership Program (BNPP) Trust Fund, (ii) the Trust Fund for Environmentally and Socially Sustainable Development (TFESSD) funded by Finland and Norway, (iii) the Norwegian Trust Fund (NTF) and (iv) the Trust Fund for Integrated Land and Water Management for Adaptation to Climate Variability and Change (ILWAC) funded by Denmark. We gratefully acknowledge the NBA Observatory for making hydro-meteorological and runoff data available, as well as providing valuable comments and directions for the study. We extend our thanks to Dr. Amal Talbi for managing this CRA study for the World Bank, and to the World Bank's Water Partnership

Program (WPP) and Water Unit for organizing the special session S3 of HydroPredict2012 (Vienna, September 2012) on 'Choosing Models for Resilient Water Resources Management' (Grijsen et. al, 2013), and for giving permission to present the results of this CRA work in this book.

Author details

J.G. Grijsen[1], A. Tarhule[2], C. Brown[3], Y.B. Ghile[4], Ü. Taner[5], A. Talbi-Jordan[6], H. N. Doffou[7], A. Guero[7], R. Y. Dessouassi[7], S. Kone[7], B. Coulibaly[7] and N. Harshadeep[8]

1 Independent Hydrology and IWRM Consultant, Arrowlake Road, Wimberley, Texas, USA

2 Department of Geography and Environmental Sustainability, University of Oklahoma, Norman, USA

3 Department of Civil and Environmental Engineering, University of Massachusetts, Amherst, USA

4 Woods Institute for the Environment, Stanford University, Stanford, USA

5 Dept. of Civil and Environmental Engineering, University of Massachusetts, Amherst, USA

6 The World Bank Middle East North Africa (MNSWA), Washington, USA

7 Niger Basin Authority/ Autorité du Bassin du Niger (ABN), Niamey, Niger

8 The World Bank, Africa Region, Washington DC, USA

References

[1] Arora, V.K., 2002: The use of the aridity index to assess climate change effect on annual runoff, Journal of Hydrology, 265, p. 164–177.

[2] Deksyos T. and T. Abebe, July 2006: Assessing the impact of climate change on the water resources of the Lake Tana sub-basin using the Watbal model. CEEPA Discussion Paper No. 30, Special Series on Climate Change and Agriculture in Africa. Centre for Environmental Economics and Policy in Africa, University of Pretoria, South Africa.

[3] Gleick, E. H., 1986: Methods for Evaluating the Regional Hydrological Impact of Global Climatic Changes, J. of Hydrology 88, 97-116.

[4]   Gleick, E. H., 1987: Regional Hydrologic Consequences of Increases of Atmospheric CO_2 and Other Trace Gases, Climatic Change 110, 137-161.

[5]   Grijsen, J.G. and C. Brown, 2013: Climate elasticity of runoff and climate change impacts in the Niger River Basin; Niger River Basin Climate Risk Assessment, a joint initiative of the Niger Basin Authority (NBA) and the World Bank *(not yet disclosed)*.

[6]   Grijsen, J.G., C. Brown, A. Tarhule: Climate Risk Assessment for Water Resources Development in the Niger River Basin; HydroPredict'2012 - International Conference on Predictions for Hydrology, Ecology and WRM, Special Session S3 – Choosing models for resilient water resources management, Vienna; World Bank WET/WPP/TWIWA publication *(in print)*.

[7]   Hirabayashi, Yukiko, Shinjiro Kanae, Ken Motoya, Kooiti Masuda and Petra Doll, 2008: A 59-year (1948-2006) global near-surface meteorological data set for land surface models. Part I: Development of daily forcing and assessment of precipitation intensity. Hydrological Research letters 2, 36-40.

[8]   Karl, T.R. and W.E. Riebsame, 1989: The impact of decadal fluctuations in mean precipitation and temperature on runoff: A sensitivity study over the United States, Climatic Change 15: 423-447.

[9]   Legates, D., H. Lins, H., and G. McCabe, 2005: Comments on "Evidence for global runoff increase related to climate warming" by Labat et al., Advances in Water Resources, 28, 1310–1315, 2005.

[10] Pike, J. G., 1964: The estimation of annual runoff from meteorological data in a tropical climate. Journal of Hydrology, Amsterdam, 2, 116–123.

[11] Risbey, J. S., and Entekhabi, D, 1996: Observed Sacramento basin streamflow response to precipitation and temperature changes and its relevance to climate impact studies, Journal of Hydrology.

[12] SNC Lavalin International, February 2007: Strategic/Sectoral, Social and Environmental Assessment of Power Development Options in the Nile Equatorial Lakes Region, Final Report, Chapter 12 and Appendix K.

[13] Strzepek K. M. and A. McCluskey, July 2006: District level hydroclimatic time series and scenario analyses to assess the impacts of climate change on regional water resources and agriculture in Africa; CEEPA Discussion Paper No. 13, Special Series on Climate Change and Agriculture in Africa.

[14] Turc, L., 1954: Le Bilan d'eau des sols. Relation entre les précipitation, l'évaporation et l'écoulement. Annales Agronomique, 5, 491–595.

[15] Vogel, R.M., I. Wilson and C. Daly, May/June 1999: Regional regression models of annual stream flow for the United States, Journal of Irrigation and Drainage Engineering.

[16] Wigley, T. M. L. and E.D. Jones, 1985: Influences of Precipitation Changes and Direct CO_2 Effects on Streamflow, Nature 314, 140-152.

[17] World Bank, 2012: Project Appraisal Document on the 1st part of the 2nd phase of the Niger Basin Water Resources Development and Sustainable Ecosystems Management Project (WRD-SEM APL 2A).

Chapter 3

Climate Risk Assessment for Water Resources Development in the Niger River Basin Part I: Context and Climate Projections

J. G. Grijsen, C. Brown, A. Tarhule, Y. B. Ghile,
Ü. Taner, A. Talbi-Jordan, H. N. Doffou, A. Guero,
R. Y. Dessouassi, S. Kone, B. Coulibaly and
N. Harshadeep

Additional information is available at the end of the chapter

1. Introduction

1.1. Study area

Climate change has emerged as one of the most important challenges of the 21st century. Consequently, a plethora of studies has appeared which focus on adaptation measures for responding to the climate change threat. A cursory review of much of the adaptation literature shows, however, that most of it contains little more than generic recommendations to vaguely defined or specified threats. This situation results from the fact that generally the specific impacts of climate change in specific locations or to specific sectors remain poorly understood and under-investigated. This situation is especially acute in Africa where adaptation strategies are urgently needed. In this chapter, we present the results of a study designed to identify and quantify specific impacts of climate change to the hydro-energy, environmental, and agro-economic sectors in the Niger Basin in West Africa.

The Niger River is the second largest river in Africa by discharge volume (5,600 m^3/s at Onitsha; 1955-1991) and the third longest (4,100 km). Taking its source on the interior side of the Guinea Daro Massif at an altitude of 1068 meters, the Niger River runs off in a northeasterly direction towards the Niger Inland Delta. At Tossaye (Taoussa, Republic of Mali), the river turns southeast forming the Great Bend and flows on through Niger to Lokoja in Nigeria, where it receives its largest and most important tributary, the Benue River (Fig. 1). From there, the enlarged Niger runs directly south to empty into the Bight of Benin through a network of distributaries in Nigeria's Terminal Niger Delta. The river's unusual flow path delineates a drainage basin that is boomerang-shaped and bounded approximately by latitudes 5° N and 22° N and by longitudes 11°30' W and 15° E. The Niger Basin's total drainage area is 2,170,500 km² although 35% (770,500 km²) of the basin lies within the hyper-arid Sahara and contributes no runoff to the Niger River. The Niger Basin cuts across all the major climatic zones of West Africa, including the Guinean or Equatorial forest zone, the Transitional tropical belt, the Sudan Savanna zone, the semi-arid or Sahel savanna belt, and the desert region. Rainfall, temperature, humidity and evaporation all exhibit steep gradients as one moves hinterland from the coast (Fig. 2).

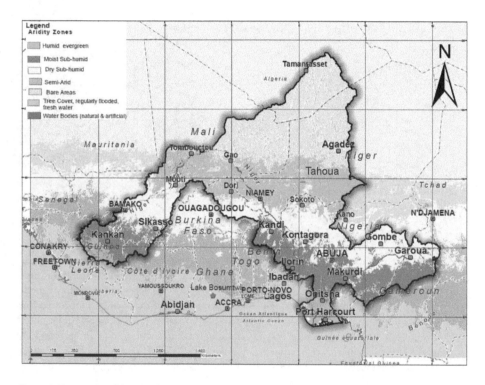

Figure 1. The Location of the Niger Basin in West Africa. Source: BRLi, 2007.

The Niger Basin is experiencing considerable pressures on ecosystem resources and with good reason: all 9 countries in the active basin rank among the bottom 40 countries on the Human Development Index (UNDP, 2010). Over 100 million people inhabit the Niger basin distributed in nine countries, namely Benin, Burkina Faso, Cameroon, Chad, Cote D'Ivoire, Guinea, Mali, Niger, and Nigeria (Fig. 1). Average population growth rate is estimated at 2.7% and the rate of urbanization at 4.3%, both numbers among the highest rates in the world (UNFPA, 2010).

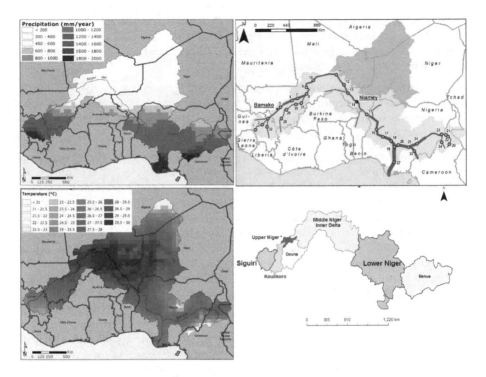

Figure 2. The climatic characteristics of the Niger Basin. The panel on the top left shows the distribution of annual average precipitation, the panel on the top right shows the basin demarcated into active and inactive parts. Also shown is the location of river gauging sites within the basin. The bottom left panel shows the distribution of annual average temperature. Finally, the bottom right panel shows the basin divided into commonly recognized sub-basins. The Upper Niger Basin includes all of the sub basins above the Inner Delta (red area) i.e. Siguiri, Koulikoro (or Sélingué) and Douna (or Bani) watersheds.

Approximately 70% of the population depends on agriculture for their livelihood and overwhelmingly on rain-fed agriculture, and 60% lack access to a developed source of water. Somewhat paradoxically, the basins' environmental and water resources are currently grossly underutilized. The irrigation potential is estimated at 2.5 million hectares (ha) but less than 0.3 million ha have been developed; hydropower generation potential is estimated at 30,000 gigawatt hours per year but only 6000 gigawatt hours have been developed; and of the basin's 3,000 km of navigation potential, only 600 km is currently utilized (The World Bank, 2005).

Recognizing the development imperatives, the governments of the member states negotiated and adopted in 2002 a shared vision for development, titled the Sustainable Development Action Plan (SDAP) Under the umbrella of the Niger Basin authority (NBA). With expected total investments of US$ 8 billion over the period 2005 – 2025, the SDAP aims to (i) provide new social and economic opportunities for more than 100 million people living in the Niger Basin, (ii) contribute to capacity building in integrated water resources management, and (iii) protect natural resources and ecosystems (BRLi, 2007). A major part of SDAP (80%) focuses on the construction of three new dams[1] on the Upper and Middle Niger with associated irrigated agriculture and hydro-energy generation, along with the rehabilitation of existing dams[2], and improved navigation and water supply.

As part of SDAP implementation, the Executive Secretariat of the NBA and the World Bank (WB) have jointly implemented a Climate Risk Assessment (CRA) initiative for the NRB. This chapter presents a robust methodology for assessing climate change impacts on the Sustainable Development Action Plan (SDAP).

The CRA initiative is prudent because of a history of marked climate variability with significant socio-economic and environmental impacts. Archival records show evidence of a regional, multi-decadal drought between 1738 and 1756, centered on the Great Bend of the River Niger but extending far south into the coastal areas of present day Ghana. Famine during this period killed half of the population of Timbuktu and other areas of the Niger Bend (Curtin, 1975). In contrast, most of the 1780s experienced excessive rainfall, as far North as Agadez in Niger. More recently, since 1970, the semi-arid Soudan and Sahel savanna zones within the region have witnessed the most dramatic example of inter-decadal climatic variability ever measured quantitatively on the planet since instrumental records have been kept (Rasmusson and Arkin, 1993; Hulme, 2001, p.20; Lebel et al., 2003). Average annual rainfall amounts fell by between 25% and 40%, depending on the location, and persisted over the next two decades, prompting some researchers to hypothesize about whether such decline represented evidence of climate change (Demarée, 1990). Widespread environmental desiccation driven by diminished rainfall and land use practices prompted routine discussions of the "expanding" or "encroaching" Sahara (Lamprey, 1975; Eckholm and Brown, 1977; Los Angeles Times, 1988).

Within this historical context, this CRA is urgent given the expected future impacts of climate change in this region (e.g. IPCC, 2007). Because of the substantial uncertainty in climate projections from the current Global Circulation Models (GCMs), it is difficult to estimate what the future climate is likely to be. This CRA thus aims to better understand the dynamics of the future climate over the NRB, and to assess its potential impacts on water resources, energy, navigation, agriculture and environment, as well as possible consequences for the design of existing and planned infrastructure. Finally, it assesses also the potential negative impacts on the Economic Internal Rate of Return (EIRR) of the SDAP. This is essential for assisting decision makers and stakeholders to better manage their resources, prepare for extreme hydrological hazards, and enhance development planning in the NRB. The CRA assessed the risks of climate change to the water resources and associated development sectors of the Niger Basin in the near (2030), mid (2050), and distant future (2070).

The study is presented in two separate chapters. This chapter introduces the study area, followed by an overview of the study approach and methodology, including the operational definition of risks adopted. Next, the impacts of potential changes in runoff on the SDAP -

---

1 Fomi dam in Guinea, Taoussa dam in Mali and Kandadji dam in Niger, referred hereafter as the FO-TA-KD development scenario.

2 Kainji and Jebba dams on the main Niger River in Nigeria, and Lagdo dam on the Benue River in Cameroon

based on an existing water resources allocation model for the basin - are presented (vulnerability analysis). Finally, the adopted approach to projection of climate changes is described. The next chapter discusses the methodology used in estimating the runoff response to climate change, assess climate risks for key water related sectors, as well as the economic internal rate of return.

2. Overview of approach and methodological framework

An essential step for climate related adaptation strategies is to estimate the risk (probability) of exceedance of critical impact levels for various impacted sectors. Such probabilities provide better insights about the likelihood that a specific climate change impact may occur, contributing to informed decisions on investments and adaptations. For this reason, we adopted a risk-based analysis framework. The premise of the study is that estimates of the plausibility of climate risks will help to develop a conceptual framework for adaptation strategies that increase the resilience and robustness of the SDAP investment plan in the NRB.

The adopted methodology is as follows (Ghile et. al, 2013):

i.   The impacts of changes in runoff on SDAP performance indicators were estimated using an existing Niger River Basin water resources system model (MIKEBASIN) by parametrically varying the hydrological conditions for the period 1966/67 – 1988/89 (vulnerability analysis). The runoff elasticities of selected performance indicators[3] were then determined (Brown and Grijsen, 2013).

ii.   Based on the Intergovernmental Panel on Climate Change (IPCC) A1B scenario, we estimated the probability distributions of changes in precipitation and temperature, using a bias corrected ensemble of 38 Global Climate Model (GCM) projections of future climate for the 21st century.

iii.   Using a variety of hydrologic models and methods, we determined the response of runoff to climate changes (defined as climate elasticity[4] of runoff) and translated the probability distributions of future changes in precipitation and temperature into probability distributions of future mean runoff (Grijsen and Brown, 2013).

iv.   Finally, we estimated the probabilities of changes in SDAP performance based on impacts of runoff changes and the assessed probability distributions of changes in runoff.

---

3 The runoff elasticity of a performance indicator (e.g. hydro-energy production) defines the response (as a multiplier) of the indicator to changes in runoff; for example, a runoff elasticity of hydro-energy of 1.2 indicates that a 10% decrease in runoff would cause a 12% decrease in generated hydro-energy.

4 Climate elasticity of runoff defines the response (as a multiplier) of runoff to changes in precipitation and temperature; for example, a precipitation elasticity of runoff of 2.5 indicates that a 10% decrease in rainfall would cause a 25% decrease in runoff. Similarly, a temperature elasticity of runoff of -0.5 indicates that an increase in temperature of 10% (e.g. from 26 to 28.60C) would cause a decrease in runoff of 5%.

2.1. Defining climatic risk in the Niger Basin

In this study climate risk is defined on the basis of changes in annual runoff as a function of projected precipitation and temperature changes; it does not take into account other sources of risk such as vulnerability of specific sectors and future socio-economic changes, nor does it account for changes in the variability of runoff within the year. As mentioned previously, we used an ensemble of 38 climate model runs to account for model error and natural climate variability, and also to obtain the full range of possible climate futures and associated runoff. Each run was treated as equally plausible. Using these runs the probability distributions of future runoff regimes, centered on the years 2030, 2050 and 2070 were fitted to estimate the plausibility of climate risks. To assess the severity and probability of climate-imposed risks in the basin, the generated future mean flows from the 38 climate projections were used to estimate risk levels for key performance metrics. In consultation with basin stakeholders in a CRA workshop in Ouagadougou (2010), reductions of more than 20% from the baseline performance were considered to be significant impacts. With this guidance, we identified five risk levels ranging from "Mild risk" to "Extreme risk" (Fig. 3), based on a choice of percentage shifts in performance relative to baseline operations. In this study the baseline refers to a development scenario in which three dams are built, along with associated irrigation infrastructure. These dams include Fomi (FO; in the Republic of Guinea), Taoussa (TA; in the Republic of Mali) and Kandadji (KD; in the Republic of Niger). Thus, the baseline scenario is interchangeably referred to as the FO-TA-KD scenario. We focused on five sectors and key performance metrics, as identified by the Ouagadougou CRA workshop participants and shown in Table 1.

Risk levels

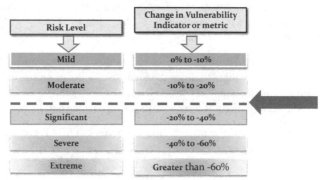

Different risk levels and categories can be chosen, and these could differ in different parts of the basin (e.g., by development zone), for major projects, and for different users

Figure 3. Risk levels based on change in vulnerability indicators. The red arrow indicates the 20% reduction in performance relative to base level beyond which impacts are considered significant.

Sector/Domain	Indicator	Baseline
Agriculture/irrigation	Incremental net irrigated area	1.5 Mha
Energy production	Total annual energy production	8,250 GWh
Navigation	Annual number of days	5 months
Inner Delta Flooding	Average annual flooded area	11,000 km^2
Environmental Flows (at Markala, Mali-Niger border, Niamey and Malanville)	Minimum flow in m^3/s	40, 75, 125 and 80 m^3/s

Table 1. Key Performance Indicators and baseline values for various sectors

2.2. Climate projections

Climate projections typically lack credibility at the spatial and temporal scales that are relevant to water resources planning, especially in tropical regions where monsoonal systems are poorly represented and inter-annual variability is poorly reproduced. While a variety of downscaling approaches are available which change the resolution of climate projections, they do little to improve their credibility. In this analysis we have attempted to address these concerns in two ways. First, as described above, a bottom-up (decision-scaling) approach to assessing risks was adopted (Brown et. al, 2012), which focuses primarily on the vulnerability of the water resources system performance to runoff changes due to climate change. As a result, the climate projections are placed in the context of risks to investments rather than as credible projections of the future. Second, the climate projections were not downscaled to higher resolution; instead, the hydrologic response of the basin was "up-scaled" to better match the resolution at which climate projections have more credibility, i.e. at the basin scale. These approaches represent an attempt to make the best use of the highly uncertain climate information described further below.

Precipitation projections for West Africa vary widely, such that GCMs even lack agreement on the direction of future changes in precipitation (Figs. 4 - 5). Therefore, we analyzed precipitation and temperature projections from an ensemble of 38 GCM simulations[5] for the 21st century, to capture a quantitative global assessment of climate-change-driven risks for the key metrics identified in the NRB. The projections were driven by the median A1B CO$_2$ emission scenario (Nakicenovic and Swart, 2000). To correct for systematic biases inherent in all GCM models, we employed the quantile mapping method - as described in Wood (2002), Wood (2004) and Segui (2010), and trained it on observed and simulated data for the 20th century. It was initially assessed that the estimation of climate risks over the entire NRB would be the most skillful scale for projections. This analysis was done for the periods 2016-2045, 2036-2065 and 2056-2085, considered representative respectively for 2030, 2050 and 2070.

We have also used climate change projections available for Guinea and Nigeria from the climate wizard website[6] (climatewizard.org) and the WBG's Climate Change Knowledge

---

5 Source: Data library of the International Research Institute for Climate and Society (IRI).

Portal (Climate Portal)- as being representative for respectively the Upper Niger Basin and the Lower Niger and Benue Basins - to determine whether climate change impacts in the Upper Niger (Guinea) and Benue (mainly Nigeria) Basins could be more or less severe than the average impacts across the Niger Basin. The Upper Niger and Benue Basins constitute the main water sources ('water towers') of the Niger River. Results are shown in Table 2, showing a good agreement between the various sources of climate change projections.

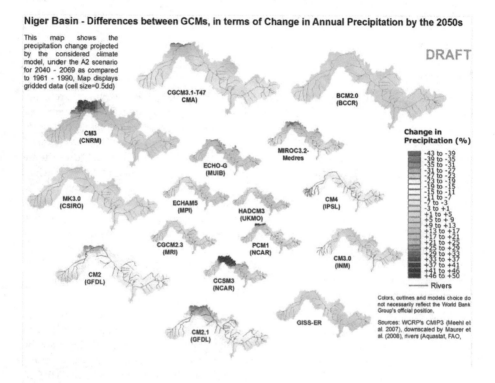

Niger Basin - Differences between GCMs, in terms of Change in Annual Precipitation by the 2050s

Figure 4. Annual precipitation changes projected for 2050 by various GCMs[7] (Sources: right lower corner)

Strzepek (2011) provides a detailed account of the studies and hydrological modeling under-pinning the results displayed in the Climate Portal. To generate a high-resolution understand-

6 University of Washington and the Nature Conservancy (2009); Data source: Global Climate Model (GCM) output, from the World Climate Research Program's (WCRP) Coupled Model Inter-comparison Project phase 3 (CMIP3) multi-model dataset (Meehl et al., 2007), was downscaled (as per Maurer et al., 2009) using the bias-correction/spatial downscaling method of Wood et al. (2004) to a 0.5 degree grid, based on the 1950-1999 gridded observations of Adam and Lettenmaier (2003).

7 The largest changes are projected for regions with low rainfall, which has small implications in absolute terms.

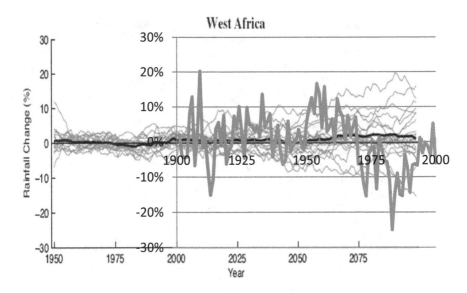

Figure 5. Ensemble of regional climate projections for West Africa (Giannini et al., 2008); the black line is the ensemble mean (A1B); the bold blue line reflects the historical 20th century rainfall variability for the Niger River Basin

ing of possible risks, the analyses examined relative changes in several hydrological indicators (including *inter alia* annual precipitation, temperature and mean runoff, high flow, low flow and potential evapotranspiration (E_0)) from the historical baseline (1961 to 1999) to the 2030s and 2050s, for a range of 56 GCM and emission scenario combinations. Projected changes in runoff and E_0 were simulated with the CLIRUN-II hydrologic model II (Strzepek and Fant, 2010), an extension of the WatBal water balance model (Yates, 1996). The results provide an understanding of the range of potential consequences of climate change on water resources at the country and basin scale. The results are claimed to be suitable for use as inputs to screening-level analyses of the impact of climate change on water resources dependent investments, for the 2050 investment horizon. Results for the various sources of climate projections agree very well with the results obtained under the present NRB-CRA study on the basis of 38 GCM model runs. Differences between impacts for the two countries are marginal. The Climate Portal shows potential evapotranspiration to increase by nearly 5% in 2050, or a climate sensitivity of E_0 of 2.3% per ^{0}C at an average annual basin temperature of 26.6 ^{0}C.

The results show that a slightly larger number of models project increases in precipitation than those projecting decreases, while all the models project increases in temperature. Fig. 6 shows quartiles of rainfall and temperature projections (0, 25, 50, 75 and 100% points), for percentage changes relative to the 20th century averages. The graphs show variations of the 30-yr average across the 38 GCM model runs, for 2030, 2050 and 2070, interpreted as the variance of the projected future mean runoff.

Variable	Min	20%	Mean	80%	Max	St. dev.
		Climatewizard.org				
Guinea						
Temperature (°C)	1.8	2.0	2.3	2.8	3.0	
Precipitation (%)	-20.0	-6.0	0.0	6.0	10.0	
Nigeria						
Temperature (°C)	1.5	1.8	2.1	2.5	2.8	
Precipitation (%)	-15.0	-4.0	2.0	10.0	15.0	
WB Climate Change Knowledge Portal						
Guinea						
Temperature (°C)	1.2	1.8	2.1	2.6	3.0	0.5
Precipitation (%)	-12.2	-5.2	0.5	5.6	12.9	6.8
Annual runoff (%)	-23.8	-13.5	-0.3	12.0	38.7	16.5
Annual PET (%)	0.7	3.9	5.0	6.7	8.1	1.7
Nigeria						
Temperature (°C)	1.2	1.6	2.0	2.4	2.7	0.4
Precipitation (%)	-13.4	-4.4	1.2	7.0	10.9	6.4
Annual runoff (%)	-31.0	-11.3	-0.2	15.1	29.9	17.0
Annual PET (%)	1.5	3.7	4.6	6.1	7.4	1.4
Projections 38 GCM model runs for Niger River Basin						
Temperature (°C)	1.2	1.6	2.1	2.6	2.9	0.5
Precipitation (%)	-5.8	-3.5	1.4	4.5	13.7	4.5
Annual runoff (%)	-19.5	-13.2	-1.9	4.7	32.3	10.9
Annual PET (%)	2.6	3.6	4.7	5.8	6.7	1.1

Table 2. Summary of projected climate changes from various sources (2050; A1B)

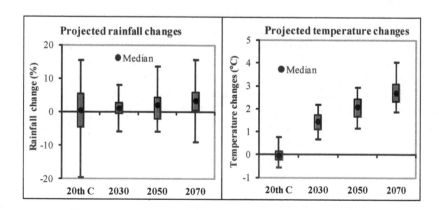

Figure 6. Quantiles for rainfall and temperature changes[8] for the 20th century and projections for 2030, 2050 and 2070

---

8 The average baseline temperature for the 20th century is 26.60C

Fig. 7 depicts annual rainfall and temperature projections for the 21st century averaged for each year over all GCM runs. It shows hardly any upward trend in precipitation till 2050 and 3% by 2070; 34 (90%) of the 38 model runs project by 2050 changes between -6% and +7%. Temperatures show a steady increase of 0.03^0C per year (2.9 ^{0}C over 100 years); on average 2.1^0C (8%) by 2050 and 2.7^0C by 2070 (10%). Projected temperature increases vary between $+1.0^0$C and $+3.0^0$C by 2050. The primary impact of increased temperatures by 2050 is an increase of the potential evapotranspiration by about 5% for a 2.1^0C increase in temperature (Grijsen, 2013), and a similar increase in gross irrigation requirements. Catchment runoff is generally also reduced due to increased evaporation, as discussed in Part II. The normal distribution has been shown to apply well to the 30-year average projections of precipitation and temperature for 2050 (Fig.8).

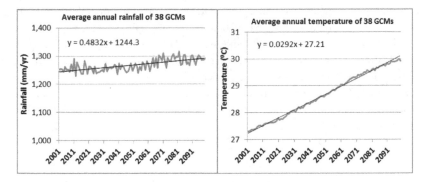

Figure 7. Annual rainfall and temperature projections for the 21st century averaged over all GCMs runs

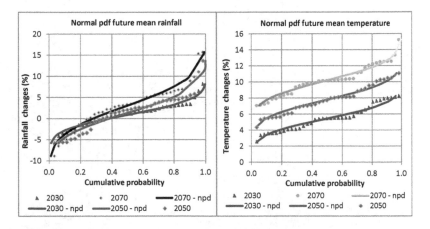

Figure 8. Normal distribution of projected average precipitation and temperature for 2030 – 2070

2.3. Assessing impacts of potential runoff changes on key SDAP performance indicators

In 2007, a Mike-Basin water resources system model was developed for the NRB to study the impact of various SDAP development scenarios on the water resources of the Basin (BRLi, 2007). We have adopted this model to simulate the water availability, water demands and water abstractions from the river and reservoir system within the NRB. The Mike Basin model is designed to simulate and assess the balance between the abstractions of water for irrigation and domestic and industrial water supply on the one hand, and energy production, flooding, navigation and environmental constraints (required minimum flows) at any location of interest on the other hand. The model requires input in the form of hydrological data (including streamflow, rainfall and potential evapotranspiration), reservoir characteristics and operating rules, minimum environmental flows, water demand series for domestic needs and irrigation, hydro-electric power generation parameters, and multiple socio-economic parameters (for irrigated agriculture, hydro-energy, navigation, environmental benefits and cost, etc.). The NRB has been divided into 60 sub-catchments, with 420 river nodes, 21 existing and 4 planned dams, 10 hydropower plants, and 93 water abstraction points for various uses (Fig. 9). The model was calibrated and verified with historical hydro-meteorological data for the period 1966/67 – 1988/89. This period contains some years with above average flow conditions during the 1960s, but is dominated by the drought conditions during the 1970s and 1980s, with below average flow conditions as shown in Fig. 10.

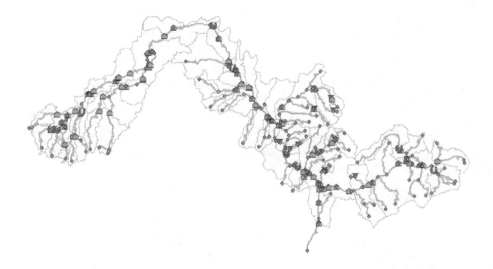

Figure 9. Schematization of the Mike Basin model for the Niger Basin.

Climate Risk Assessment for Water Resources Development in the Niger River Basin
Part I: Context and Climate Projections

65

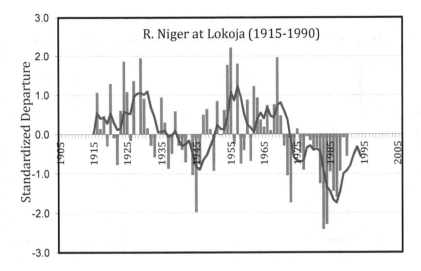

Figure 10. Standardized annual flow data for Lokoja at the confluence of the main Niger and Benue Rivers.

The relationships between relative changes in runoff and relative changes in specific performance criteria at basin level were derived from numerous Mike-Basin runs for the FO-TA-KD development scenario, for parametrically induced variations in runoff between -30% and + 10% of the 20th century baseline conditions, and water demands with a 5% increase compared to the 20th century baseline (Table 3). The average projected temperature increase by 2050 is 2.1ºC. Based on the Penman-Monteith and modified Hargreaves methods this is estimated to cause an increase of 5% in potential evapotranspiration and crop water demand (Grijsen, 2013). Fig. 11 shows the runoff changes required to cause specific changes in selected performance indicators, such as -10%, -20% etc. The 1/5 yrs (20%) performance indicators are slightly more sensitive to runoff changes than the average performances.

Under the FO-TA-KD development scenario irrigated agriculture will receive a major boost, increasing the average total harvested area by 440%. The average basin wide hydro-energy production is estimated to increase due to SDAP by 13% (including the contributions of multiple run-of-river hydro-plants) and the average navigable months in the NRB are estimated to decrease by about 20%. Flooding of the Inner Delta will face a decrease of about 10% due to SDAP. However, hydro-energy production at Kainji and Jebba dams on the main stem Niger River in Nigeria will face a decrease of nearly 35%, caused by the large incremental upstream irrigation water withdrawals. Minimum flows in the Middle Niger will be significantly improved, while minimum flows through the Inner Delta will remain well above the required 40m³/s.

Table 3 demonstrates the sensitivity of performance indicators for changes in runoff. Significant reductions of more than 20% of the baseline performance generally occur for hydro-energy production, flooding of the Inner Delta and navigation when the long-term average basin

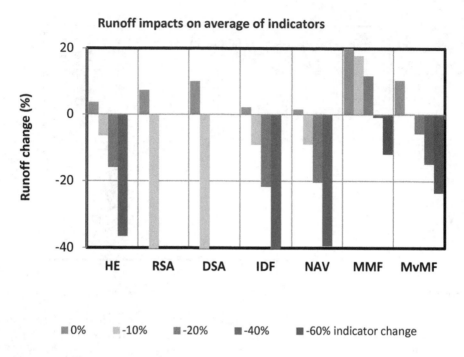

Figure 11. Runoff changes required to generate specific changes in performance indicators

runoff decreases by 15 to 20%, or more. Such long-term average decrease would represent a major increase in the aridity of the basin[9]. This sensitivity to runoff changes is attributed to the fact that energy production is not prioritized in the applicable water allocation rules under the existing Water Charter for the Niger Basin. The Charter assigns the highest priority to irrigation and domestic water supply to secure food production and alleviate poverty. Thus, irrigated agricultural is the least sensitive to climate change impacts on basin runoff. Moreover, during the rainy season water availability remains abundant and secures in most years full replen-ishment of reservoirs for dry season irrigation, even after significant flow reductions. Mini-mum flows at Markala (upstream of the Inner Delta) and in the Middle Niger (Table 3) are extremely sensitive to changes in basin runoff and cannot be maintained even under a 10% reduction in runoff. This is reflected in the runoff elasticity of performance indicators (last column of Table 3), defined as the ratio between relative changes in an indicator and relative changes in runoff[10]. Runoff elasticities of performance indicators can thus be divided in three groups, i.e. (i) irrigated agriculture with a runoff elasticity of 0.1 to 0.2 (non-sensitive); (ii) hydro-energy, navigation and Inland Delta flooding with runoff elasticities of 1.0 to 1.3 (sensitive), and (iii) minimum flows with elasticities of 2 to 4 (extremely sensitive). The dry

---

9 It is noted that the droughts of the 1970s and 1980s are already reflected in the 2005 baseline flow regime.

Climate Risk Assessment for Water Resources Development in the Niger River Basin
Part I: Context and Climate Projections

67

season minimum flows are also very sensitive to increased dry season irrigation demands due to climate change.

Performance Metrics	Probability of non-exceedance	Reference 2005 (SA)	Impacts FO-TA-KD	20thC+5% water demands; FO-TA-KD						Average runoff elasticity
				20thC-R	+10% R	0% R	-10% R	-20% R	-30% R	
		Value	% change	Value	Percentage changes (%)					
Hydro-energy										
Basin energy	1/2(50%)	7,376	12.6	8,303	5.2	-3.0	-13.9	-24.3	-33.8	1.0
(GWh)	1/5(20%)	5,771	3.4	5,969	8.9	-3.5	-16.0	-28.5	-38.7	1.3
Kainji/Jebba	1/2(50%)	5,007	-32.5	3,381	11.1	-5.8	-23.0	-37.3	-48.2	1.7
	1/5(20%)	3,743	-43.9	2,100	9.8	-5.1	-21.3	-38.0	-50.3	1.7
Irrigated Agriculture										
Total irrigation	mean	228,138	435	1,220,591	0.1	-0.3	-0.9	-1.8	-3.6	0.1
RS(ha)	1/5(20%)	228,138	424	1,194,537	-0.1	-0.5	-2.1	-4.2	-8.1	0.2
Total irrigation	mean	111,744	471	637,537	-0.4	-0.8	-1.2	-1.5	-2.2	0.1
DS(ha)	1/5(20%)	105,130	500	630,890	-0.7	-0.9	-1.4	-5.5	-15.7	0.3
Navigation for various reaches (average number of days)										
Average	Large boats	171	-20.9	135	7.9	-1.4	-10.9	-19.7	-30.2	1.0
Flooding (km²)	mean	12,117	-9.7	10,940	5.4	-1.5	-10.9	-18.7	-28.6	0.9
Inland Delta	1/5(20%)	10,342	-14.1	8,887	7.1	-1.6	-13.9	-24.8	-37.3	1.2
Sustenance of 10-day average minimum flows (m³/s)										
Markala	1/2(50%)	70	-13	61	-25.0	-38.6	-54.7	-83.5	-100.0	4
	1/5(20%)	51	-2	50	-31.1	-40.6	-69.8	-99.6	-100.0	5
Mali-Niger	1/2(50%)	60	33	80	-2.7	-12.3	-32.6	-65.2	-89.0	3
border	1/5(20%)	12	547	80	-15.9	-32.3	-63.9	-82.1	-97.4	4
Niamey	1/2(50%)	55	78	99	2.0	-9.2	-16.9	-37.5	-80.9	2
	1/5(20%)	9	818	85	-0.9	-11.3	-33.6	-73.7	-97.0	3.5
Malanville	1/2(50%)	68	35	91	-0.3	-10.3	-27.2	-53.9	-71.6	2.5
	1/5(20%)	4	2,035	77	-2.1	-21.9	-55.8	-83.5	-98.2	4

Table 3. Performance indicator matrix for the FO-TA-KD scenario (-10% R means a 10% reduction of runoff)

3. Conclusions

Effectively managing climate change will constitute one of the major challenges of the 21st. This challenge will be especially acute in those regions of the world already prone to adverse climatic variability and where the financial and technical resources for dealing with climate change are lacking. The River Niger Basin in West Africa fits this profile. However, in collaboration with several donor and investment partners, the member states of the Niger

---

10 A runoff elasticity of +1.5 indicates that a 10% decrease in runoff causes a 15% decrease in performance.

Basin under the aegis of the Niger Basin Authority are implementing a novel climate risk assessment study in an effort to climate-proof new investments. This chapter described a part of that study, including an over view of the methodology, a description of the climate change estimation procedure and an assessment of the impact of potential runoff changes on key SDAP performance indicators. The key conclusions of the study will be summarized in the next chapter (part II).

Climate change confronts decision makers with deep uncertainty, where decisions can go wrong if decision makers assume that risks are well-characterized while in reality uncertainties could be underestimated. Such conditions require robust decision making to inform good decisions by identifying system vulnerabilities and assessing alternatives for ameliorating those vulnerabilities. A water resources system model and the Performance Indicator Matrix (Table 3) prepared on the basis of a large number of model simulations, provide a powerful tool to identify early on in the CRA process, even before climate projections have been processed, system vulnerabilities and future climate change conditions where the proposed SDAP plan may fail to meet its goals. These conditions can then be used to early identify potential actions to address vulnerabilities and evaluate tradeoffs among them. This approach, which focuses on water resources system vulnerability rather than on climate change projections, is considered crucial in view of the large uncertainties embedded in the presently available climate projections for the 21st century.

Acknowledgements

This study was done as part of the Climate Risk Assessment (CRA) for the Niger Basin, a joint initiative of the Niger Basin Authority (NBA) and the World Bank to assess the risks from climate change to the performance of NBA's Sustainable Development Action Plan (SDAP) for the Niger Basin. The aim of this initiative is to build resilience to climate risks into the SDAP. The Niger CRA study is supported by grants from (i) the Bank-Netherlands Partnership Program (BNPP) Trust Fund, (ii) the Trust Fund for Environmentally and Socially Sustainable Development (TFESSD) funded by Finland and Norway, (iii) the Norwegian Trust Fund (NTF) and (iv) the Trust Fund for Integrated Land and Water Management for Adaptation to Climate Variability and Change (ILWAC) funded by Denmark. We gratefully acknowledge the NBA Observatory for making hydro-meteorological and runoff data available, as well as providing valuable comments and directions for the study. We extend our thanks to Dr. Amal Talbi for managing this CRA study for the World Bank, and to the World Bank's Water Partnership Program (WPP) and Water Unit for organizing the special session S3 of HydroPredict2012 (Vienna, September 2012) on 'Choosing Models for Resilient Water Resources Management' (Grijsen et. al, 2013), and for giving permission to present the results of this CRA work in this book. We thank the World Bank for permission to report this work here.

Author details

J. G. Grijsen[1], C. Brown[2], A. Tarhule[3*], Y. B. Ghile[4], Ü. Taner[5], A. Talbi-Jordan[6], H. N. Doffou[7], A. Guero[7], R. Y. Dessouassi[7], S. Kone[7], B. Coulibaly[7] and N. Harshadeep[8]

*Address all correspondence to: atarhule@ou.edu

1 Independent Hydrology and IWRM Consultant, Texas, USA

2 Department of Civil and Environmental Engineering, University of Massachusetts, Amherst, USA

3 Department of Geography and Environmental Sustainability, University of Oklahoma, Norman, USA

4 Woods Institute for the Environment, Stanford University, Stanford, USA

5 Dept. of Civil and Environmental Engineering, University of Massachusetts, Amherst, USA

6 The World Bank Middle East North Africa (MNSWA), Washington, USA

7 Niger Basin Authority/ Autorité du Bassin du Niger (ABN), Niamey, Niger

8 The World Bank, Africa Region, Washington DC, USA

References

[1] BRLi et alJanuary (2007). Evaluation des prélèvements et des besoins en eau pour le modèle de simulation du basin du Niger, Final Report.

[2] BRLi et alJuly (2007). Elaboration du Plan d'Action de Développement Durable du Bassin du Niger, Phase II : Schéma Directeur d'Aménagement et de Gestion.

[3] BRLi et alSeptember (2007). Establishment of a Water Management Model for the Niger River Basin, Final Report.

[4] Brown, C, & Grijsen, J. G. (2013). Sensitivity of SDAP performance to changing water availability and demands; the Niger River Basin Climate Risk Assessment, a joint initiative of the Niger Basin Authority and the World Bank (*not yet disclosed*).

[5] Brown, C, Ghile, Y, Laverty, M, & Li, K. (2012). Decision scaling: linking bottom-up vulnerability analysis with climate projections in the water sector, *Water Resources Research*, 48(9), 1-12

[6]   Curtin, P. D. (1975). Economic Change in Pre-Colonial Africa: Supplementary Evidence, University of Wisconsin Press, Madison.

[7]   Demaréee, G. R. (1990). An indication of climatic change as seen from the rainfall data of a Mauritanian station, Theoretical and Applied Climatology, , 42, 139-147.

[8]   Eckholm, E, & Brown, L. R. (1977). Spreading deserts-The hand of man, World watch Paper Washington D.C.(13)

[9]   Giannini, A, Biasutti, M, Held, I. M, & Sobel, A. H. (2008). A global perspective on African climate, Climatic Change , 90(4), 359-383.

[10] Ghile, Y. B, Taner, M. Ü, Brown, C, & Grijsen, J. G. (2013). Bottom-up Climate Risk Assessment of Infrastructure Investment in the Niger River Basin, Climate Change (under review)

[11] Grijsen, J. G. (2013). Future Water Demands in the Niger Basin, the Niger River Basin Climate Risk Assessment, a joint initiative of the Niger Basin Authority (NBA) and the World Bank (*not yet disclosed*).

[12] Grijsen, J. G, & Brown, C. (2013). Climate elasticity of runoff and climate change impacts in the Niger River Basin; the Niger River Basin Climate Risk Assessment, a joint initiative of the Niger Basin Authority (NBA) and the World Bank (*not yet disclosed*).

[13] Grijsen, J. G, Brown, C, & Tarhule, A. Climate Risk Assessment for Water Resources Development in the Niger River Basin; HydroPredict'(2012). International Conference on Predictions for Hydrology, Ecology and WRM, Special Session S3- Choosing models for resilient water resources management, Vienna; World Bank WET/WPP/ TWIWA publication (*in print*)., 2012.

[14] Hulme, M. (2001). Climatic perspectives on Sahelian desiccation: 1973-1998. Global Environmental Change , 11, 19-29.

[15] IPCCKundzewicz, Z.W., L.J. Mata, N.W. Arnell, P. Döll, P. Kabat, B. Jiménez, K.A. Miller, T. Oki, Z. Sen and I.A. Shiklomanov, (2007). Freshwater resources and their management. Climate Change 2007: Impacts, Adaptation and Vulnerability. Contribution of Working Group II (Chapter 3) to the Fourth Assessment Report of the Intergovernmental Panel on Climate Change, M.L. Parry, O.F. Canziani, J.P. Palutikof, P.J. van der Linden and C.E. Hanson, Eds., Cambridge University Press, Cambridge, UK, , 173-210.

[16] Lamprey, H. F. (1975). Report on the desert encroachment reconnaissance in northern Sudan, 21 Oct. to 10 Nov., 1975; UNESCO/UNEP.

[17] Lebel, T, Redelsperger, J. L, & Thorncroft, C. (2003). African Monsoon Multidisciplinary Analysis (AMMA): an international research project and field campaign; CLIVAR Exchanges , 8, 52-54.

[18] Los Angeles Times(1988). Desert Encroachment: Fabled Oasis of Timbuktu Is Drying Up; byline by Scott Kraft, March 08, 1988; available online at http://articles.latimes.com/1988-03-08/news/mn-608_1_desert-encroachment;last accessed September 3, 2012.

[19] Maurer, E. P, Adam, J. C, & Wood, A. W. (2009). Climate Model based consensus on the hydrologic impacts of climate change to the Rio Lempa basin of Central America, Hydrology and Earth System Sciences, 13, 183-194.

[20] Meehl, G. A, Covey, C, Delworth, T, Latif, M, Mcavaney, B, Mitchell, J. F. B, Stouffer, R. J, & Taylor, K. E. (2007). The WCRP CMIP3 multi-model dataset: A new era in climate change research, Bulletin of the American Meteorological Society, , 88, 1383-1394.

[21] Nakicenovic and Swart(2000). Special Report on Emissions Scenarios, Cambridge University Press, Cambridge, UK. Available at http://www.grida.no/climate/ipcc/emission/023.htm.

[22] Rasmusson, E. M, & Arkin, P. A. (1993). A global view of large-scale precipitation variability. Journal of Climate , 6, 1495-1522.

[23] Segui, P. Q, Ribes, A, Martin, E, Habtes, F, & Boe, J. (2010). Comparison of three downscaling methods in simulating the impact of climate change on the hydrology of Mediterranean basins, Journal of Hydrology , 383, 111-124.

[24] Strzepek, K. M, Mccluskey, A, Boehlert, B, Jacobsen, M, & Fant, C. W. IV, (2011). Climate Variability and Change: A Basin Scale Indicator Approach to Understanding the Risk to Water Resources Development and Management, Water Anchor of the World Bank Group, series Water Papers: WB Water Paper CCK Portal.

[25] Strzepek, K. M, Fant, C. W, & Water, I. V. and Climate Change: Modeling the Impact of Climate Change on Hydrology and Water Availability, University of Colorado and Massachusetts Institute of Technology.

[26] United Nations Development Program(2010). Human Development Indicators-2010 Rankings. The United Nations Development Program. Available online at http://hdr.undp.org/en/media/PRHDR10-HD1-E-rev4.pdf.

[27] United Nations Population Fund(2010). *The state of the world population: Demographic and economic indicators.* UNPFA. Available online at http://www.unfpa.org/swp.

[28] World Bank(2005). *The Niger River Basin: A vision for sustainable management.* The World Bank. Available online at http://elibrary.worldbank.org/content/book/9780821362037.

[29] Yates, D. (1996). WatBal: An integrated water balance model for climate impact assessment of river basin runoff. International Journal of Water Resources Development , 12(2), 121-139.

[30] Wood, A. W, Maurer, E. P, Kumar, A, & Lettenmaier, D. P. (2002). Long-range experimental hydrologic forecasting for the eastern United States, Journal of Geophysics
      Research 107(D20), 4429.

[31] Wood, A. W, Leung, L. R, Sridhar, V, & Lettenmaier, D. P. (2004). Hydrologic implications of dynamical and statistical approaches to downscaling climate model outputs, *Climate Change* , 62, 189-216.

Chapter 4

Global Warming —
Scientific Facts, Problems and Possible Scenarios

M.G. Ogurtsov, M. Lindholm and R. Jalkanen

Additional information is available at the end of the chapter

1. Introduction

Climate is the average pattern of weather for a particular region, which is characterized by weather statistics (temperature, pressure, humidity, speed and direction of wind etc.) averaged over time intervals generally longer than 30 years.

Climatology is a branch of atmospheric sciences concerned with the study of climates of the Earth and analyzing the causes and practical consequences of climatic changes. Until the 19th century climatology was closely linked with meteorology. The concept of climate appeared first in ancient Greece. Aristotle (384–322 BC) wrote "Meteorologica" – the first scientific book about the atmospheric (meteorological and climatic) phenomena. The understanding of climate by Aristotle and Hippocrates (460–377 BC) remained very influential until well into the 18th century [38]. The medieval Chinese scientist Shen Kuo (1031–1095) was probably the first person who asserted that climate can change in the course of time. Enlightenment in Europe gave a new pulse to the development of both weather and climate research. The invention of meteorological devices – thermometer (Galileo in 1603), mercury barometer (Torricelli in 1643), barometer-aneroid (Leibnitz in 1700) – opened a new era in meteorology. Appreciable milestones in both meteorology and climatology were reached in the 19th century. In 1817 A. Humboldt (1769–1859) – one of the pioneers in scientific climatology – constructed the first map of global annual isotherms using the data from 57 weather stations. In 1848 H.W. Dove (1803–1879) constructed maps of the isotherms of January and July. The first isobars based on data on prevailing winds of the entire globe were constructed by Buhan in 1869. Francis Galton (1822–1911) invented the term anticyclone. Moreover, in 1896 S. Arrhenius (1859–1927) claimed that fossil fuel combustion may eventually result in a global warming. He proposed a relation between atmospheric carbon dioxide concentrations and global temperature. These among numerous other findings laid the foundation for modern climatology.

Global warming (GW) has become the most interesting problem of climatology in the second part of the 20th century. By the end of the 1980s it was finally acknowledged that global climate is warmer than during any period since 1880. Climatic modeling, including the greenhouse effect theory, started to develop intensively and the Intergovernmental Panel on Climate Change (IPCC) was founded by the United Nations Environment Programme and the World Meteorological Organization. This organization aims at assessing the scientific information of the risk of human-induced climate change and prediction of the impact of greenhouse effect according to existing climate models. The problem of global warming has also moved from the realm of scientific debates into global and local political spheres. What is the physical mechanism of GW? Does it result only from anthropogenic activity (especially the burning of fossil fuels) or do some other natural climatic phenomena contribute to the global temperature increase, too? What is the magnitude and pattern of the warming? Answers to these questions can provide us valuable information about potential climate changes in future decades and, hence, is of crucial importance for all human activity. However, in order to answer the above questions, we need detailed information about past climatic variability and its causes. Unfortunately, substantial gaps exist in our knowledge concerning the dynamics of climate variability. The available instrumental (meteorological) records are sparse and irregularly distributed. They usually cover no more than the last 100–150 years. Paleoclimatic proxy records (reconstructions from natural archives) are irreplaceable tools in filling the gaps in our knowledge about long-term climatic changes. However, the paleodata are less accurate and their reliability quite often raises serious doubts. On the other hand, modern climate models include numerous parameters, some of which are not defined adequately enough. Therefore an integrated analysis using different approaches is necessary to obtain a clearer picture of the global warming.

2. Global warming in the context of instrumental data

The average temperature of the Earth, measured by the surface weather station thermometers, has increased appreciably during the last century. The IPCC consortium reported that the global mean surface temperatures have risen by 0.74°C ± 0.18°C when estimated by a linear trend over the last 100 years [42] (Figure 1A).

According to a currently widely held view, the temperature rise is: (a) mainly a result of anthropogenic emissions of greenhouse gases (CO_2, CH_4, N_2O, halocarbons) and (b) extremely high and unprecedented in a historical context (see e.g. [42]). It is, however, evident that the instrumentally recorded temperature data are not representative enough to support solid conclusions. Even during the last few decades the weather station network covers less than 90% of the land, i.e. no more than 25% of the Earth's surface (see Figure 1B). Moreover, the scarcity of spatial coverage of these data generally increases going back in time. That is why the uncertainty of the global annual temperature increases from less then 0.1°C at the end of the 20th century to about 0.15°C at the end of the 19th century [14].

Satellite measurements of atmospheric temperature, started at the end of the 1970's, have much more dense spatial coverage. The satellite passes over most points on the Earth twice per day.

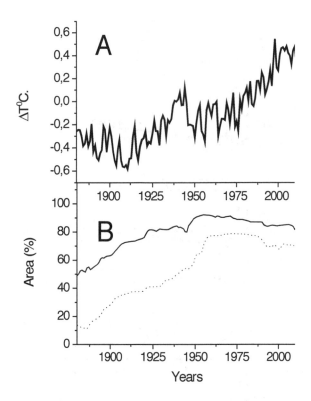

Figure 1. A – observed changes in a global average annual temperature (http://www.cru.uea.ac.uk/cru/info/warming/. B – the percent of hemispheric area located within 1200 km of a reporting weather station. Solid line – Northern Hemisphere, dotted line – Southern Hemisphere. Data were electronically scanned from http:// data.giss.nasa.gov/gistemp/station_data/#form and digitized.

Satellite-borne microwave sounders – the *microwave sounding units* (MSU) – estimate the temperature of thick layers of the atmosphere by measuring microwave thermal emissions (radiances) of oxygen molecules from a complex of emission lines near 60 GHz. By making measurements at different frequencies near 60 GHz ($\cong 1$ cm), different atmospheric layers can be sampled. Then, based on the obtained data and by means of various mathematical procedures, atmospheric temperature is calculated. Two groups of scientists – the Remote Sensing System (RSS) group and the University of Alabama (UAH) group – have analyzed the data produced by NASA (National Aeronautics and Space Administration) satellites series TIROS (Television Infrared Observing Satellites) and obtained two versions of temperature changes in the lower troposphere, i.e. at heights less than 8 km (maximum of sensitivity around 2–3 km) since the end of 1978. Figure 2 shows RSS and two UAH satellite series (versions 2007 and 2012) together with the surface thermometric data.

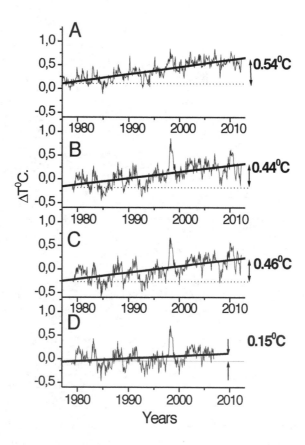

Figure 2. A – observed changes in a global average temperature (ftp://ftp.ncdc.noaa.gov/pub/data/anomalies/monthly.land_ocean.90S.90N.df_1901-2000mean.dat) ; B – RSS satellite based global lower troposphere temperature anomaly (http://www.remss.com/data/msu/); C – UAH satellite based global lower troposphere temperature anomaly (version of 2012, http://vortex.nsstc.uah.edu/public/msu/); D – UAH satellite based global lower troposphere temperature anomaly (version of 2007, http://www.ncdc.noaa.gov/oa/climate/research/msu.html).

It should be noted, however that it is not a trivial task to tie readings from different satellites together. Satellite measurements are limited by the following constraints [66]: (a) offsets in calibration between satellites; (b) drifts in satellite calibration; (c) orbital decay and drift and associated long-term changes in the time of day that the measurements are made at a particular location. Therefore it is not easy to evaluate correctly long-term trends in the satellite micro-wave-based data and the estimations often vary. The initial versions (before 2005) of UAH record (Figure 2D) had a weak trend of 0.03-0.05^{0}C /decade [21]. In 2005 the trend value changed to 0.12^{0}C /decade [22]. The current version (Figure 2C) has a trend value about 0.14^{0}C / decade, which is very close to that in RSS series (Figure2B). In spite of some past disagreement, both UAH and RSS records show that the temperature rise in the low troposphere is unlikely less than at the surface contrary to the predictions of greenhouse warming theory. Actually, physical theory and the GCMs predict that the troposphere will warm faster than the surface as the greenhouse effect takes place (see IPCC, 2007). This effect is most expressed in the tropical troposphere, because this part of the atmosphere is the most appropriate for detecting the greenhouse fingerprint. IPCC GCMs forecast a tropical tropospheric greenhouse warming increasing with altitude and reaching its maximum at ca 10 km (see Fig. 9.1 of [40]). Compar-ison of the model predicted and satellite-derived temperatures has brought rather controver-sial results that created a decades-long debate which still continues. For example, Douglass et al. [28] examined tropospheric temperature trends using 22 GCMs and arrived at a conclusion that the model results and observed temperature trends are in disagreement in the largest part of the tropical troposphere, being separated by more than twice the uncertainty of the model mean. Santer et al. [84] who used the updated and corrected observational datasets found, however, that they are within the confidence intervals of the models. Fu et al. [36]), in turn, examined the GCM-predicted and observed trends in the difference between the tropical upper troposphere and lower-middle troposphere temperatures for 1979–2010 and showed that models significantly exaggerate the trend value.

3. Global warming in the context of paleodata

Paleoclimatology is the science that studies the climatic history of the Earth using proxy records. It reconstructs and studies climatic variations prior to the beginning of the instru-mental era by means of a variety of proxy sources. Among the main data used by paleoclima-tology are: tree rings (width and density), tree height growth, concentration of stable isotope (^{18}O, ^{13}C, D) in natural archives (ice, coral and tree rings), borehole temperature measurement and contemporary written historic records (weather diaries, annals etc.). The detailed descrip-tion of these paleoindicators can be found in [3, 11, 13, 43, 78]. Here we describe shortly some of the main features of the available proxy records (Table 1).

Some proxy characteristics are summarized in Table 1: (a) maximum duration of reconstruc-tion currently achieved; (b) regions of the world where reconstructions are potentially available; (c) largest coefficient of correlation between annually resolved proxy and instru-mental data; (d) largest coefficient of correlation between decadal proxy and instrumental data; (e) general shortcomings of the proxy. Coefficients of correlation were calculated, in general, over the last 80–100 years.

Proxy variable	Maximum time span	Spatial limitations	Maximum R_I inter-annual	Maximum R_I inter-decadal	General shortcomings
Tree-ring width	7–8 millennia	Extratropical (> 30°) part of the globe or high-elevation area	0.50	0.80	Standardization methods hamper interpretation of multi-decadal and longer variability. Divergence problem.
Tree-ring density	1.5–2 millennia	Extratropical (> 30°) part of the globe or high elevation area	0.79	0.92	Standardization methods hamper interpretation of multi-decadal and longer variability. Divergence problem.
Stable isotopes in tree rings	1–2 millennia	Extratropical (> 30°) part of the globe or high elevation area	0.68	0.81	A complicated procedure of measurement
Tree-height increment	1263 years	Northern Fennoscandia	0.61	0.72	A novel approach not profoundly examined
Stable isotopes in ice	750 000 years	High-latitude (>60°) and high-elevation ice caps	0.30	0.50	A complicated procedure of measurement. Problems with dating deep layers Problems with calibration towards instrumental data.
Stable isotopes in corals	1 000 000 years	Tropic (±30°) oceans	0.41	0.66	Problems with interpre-tation of multi-decadal and longer variability (changes in water depth, nutrient supply).
Contemporary written historic records	1372 years [5]	Europe, China, Japan, Korea, Russia, Egypt	0.90	0.93	Problems with interpretation of variability longer than a human lifespan
Borehole temperature	20 000 years (usually - 500 years)	Mid-latitude (30–60° N) part of Northern Hemi-sphere, south (>0° S) part of Africa, extracon-tinental Australia	–	–	Reproduce only long-term (century-scale and longer) variability.
Ice core melt layers	600 years	High-latitude (>60°) and high-elevation ice caps where tempera-ture in summer reaches positive values	0.18	0.74	Problems with calibration towards instrumental data. Might not reproduce the full range of the temperature variability

Table 1. Characterization of different paleoproxies of temperature

Tree rings are one of the most widely used climate proxies because they can be absolutely dated annually by means of dendrochronological cross-dating method. Progress in the methodology, theory and application of dendroclimatology in the last decades of the 20th century [23, 27, 34] has helped this science to become popular, and over the last decades its methods have become key tools in the reconstruction of past temperatures in many parts of the world [11, 30, 37, 55, 57, 92]. It should also be noted that tree-ring data are generally collected from territories that are remote from areas of human activity and are less subjected to local anthropogenic impacts such as urbanization and changes in land use.

Since different paleoindicators reflect actual temperature changes in different ways the *multiproxies* – the time series, which generalize proxy sets of various types – are often used by paleoclimatology [60–61].

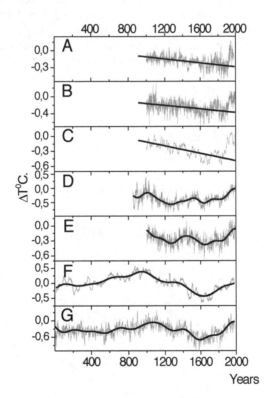

Figure 3. Reconstructions of the Northern Hemisphere temperature over the last 1–2 millennia: A) NHM – the multi-proxy of Mann et al. [54]. B) NHJ – the multiproxy of Jones et al. [43]. C) NHC – the multiproxy of Crowley and Lowery [24]. D) NHE – the tree-ring proxy of Esper et al. [30]. E) NHB – the tree-ring proxy of Briffa [11]. F) NHL – the non tree-ring proxy of Loehle [51]. G) NHMb – the multiproxy of Moberg et al. [67]. All the data sets, with the exception of the series NHL and NHMb, were recalibrated by Briffa and Osborn [10]. Grey curves represent raw data. Thick black lines represent the calculated long-term tendencies.

Based on both instrumental and paleoclimatic data, IPCC [42] claimed that the average Northern Hemisphere temperatures during the second half of the 20th century were higher than during any other 50-year period in the last 500 years with a probability >0.9 and the highest in at least the past 1300 years with a probability >0.66. Let us examine this suggestion in detail with seven reconstructions of the Northern Hemisphere temperature during the last 1000–2000 years (Figure 3). They are tree-ring proxies after Briffa [11] (NHB) and Esper et al. [30] (NHE), non tree-ring proxy after Loehle [51] (NHL) and multi-proxies after Jones et al. [43] (NHJ), Mann et al. [54] (NHM), Crowley and Lowery [24] (NHC), and Moberg et al. [67] (NHMb).

The millennial climate proxies demonstrate evidently different histories of temperature variability over the past 1000–2000 years (Figure 3). Power spectra of the temperature reconstructions demonstrate the differences even more clearly. Figure 4 shows the global wavelet spectra for the seven proxy records concerned, calculated using the Morlet basis for the years before 1900, i.e. for a time interval prior to a possible strong anthropogenic impact. Overall linear trends were preliminarily subtracted from each of the time series. One can see from Figure 4 that millennium-scale variations are present only in spectra of series NHMb and NHL. Variations with period less than 225 yrs are quite weak in the spectra of reconstructions NHM, NHJ, NHC.

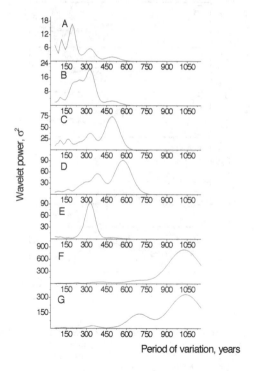

Figure 4. Global wavelet spectra of: A) NHM [54]. B) NHJ [43]. C) NHC[24]. D) NHE [30]. E) NHB [11] F) NHL [51]. G) NHMb [67]. Linear trends were subtracted prior to analyses.

A visual inspection and wavelet analysis make it possible to divide temperature reconstructions conditionally in the following groups:

a. Reconstructions NHM, NHJ and NHC (Figure 3A,B) show an obvious linear decline of mean temperature (up to 0.15–0.30⁰C) over the pre-industrial era (AD 1000–1880) and a sharp rise thereafter (the so-called hockey stick shape). Temperature in the middle of the 20th century is obviously highest in the entire millennium. According to [12] these proxies show that the Earth in the 20th century is warmer than it was in any other time period in the last millennium. We combine NHM, NHJ, and NHC records as a group, depicting a temperature pattern with a sharp rising shape – a hockey stick (IHS).

b. Reconstructions NHE and NHB (Figure 3C, D) do not show a linear trend. Instead, their long-term variability is dominated by multi-centennial cyclicities. The 20th century was warm, but the warming is not so anomalous. Here we combine NHE and NHB temperature reconstructions as a group depicting a shape of multi-centennial variability (MCV).

c. Reconstructions NHMb and NHL (Figure 3E, F) also show some linear trend in AD 1000–1880 but the temperature increase during the 20th century is not abrupt. The warming of the 20th century is comparable with that during the Medieval Warm Period (AD 800–1100), i.e. it is not unusual. Time variation with a period longer than 1000 years obviously prevails in the spectra of these records. We further call the NHMb and NHL proxies as the millennial-variability (MV) reconstructions.

It has been shown that the disagreement between the different temperature patterns cannot be fully explained by differences either in standardization techniques or in the different geographical coverage (see, for example, [72]). Thus, paleoseries of IHS, MCV and MV types can be assessed as three different clusters of seven temperature reconstructions. Bürger [19] analyzed ten temperature reconstructions by means of a more sophisticated technique and concluded that they form five clusters all of which are significantly incoherent with each other.

We have plotted two geothermal reconstructions of the global surface temperature (Figure 5) – the series by Rutherford and Mann [83] and the series by Beltrami [6]. These series are plotted together with the instrumental data [44].

Ogurtsov and Lindholm [72] showed that the borehole-based reconstructions demonstrate a history of past temperature changes conflicting with the IHS-type proxies. The IHS-type proxies have a downward linear trend during the pre-anthropogenic era (AD 1500–1880) while the corresponding trend in the global borehole temperature is evidently upward. Agreement between borehole data and MV/MCV proxies is better. The geothermal reconstructions show that the 20th century (AD 1900–2000) is the warmest period for the last 500 years. The most recent borehole-based reconstructions of Huang et al. [40], spanning the last 20 000 years, show that the average global temperature 4.5–9.0 kA before present was higher than it is today. However, the uncertainty of temperature reconstruction in such a remote past period is quite large.

The discourse about the disagreement between global paleoproxies is important because of the problem of credibility and confidence of the available temperature reconstructions, which

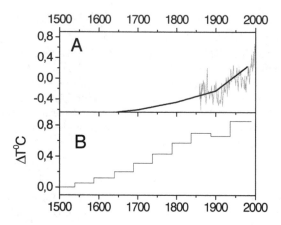

Figure 5. Change in global temperature. Grey lines – instrumental data, black lines – borehole-based reconstructions. A) Rutherford and Mann I. [83]. B) Beltrami [6]. The data from [6] were electronically scanned and digitized.

has become especially heated after the works of McIntyre and McKitrick [62-65]. These authors applied the methods of data transformation used by Mann et al. [44] to the same source data and obtained a Northern Hemisphere temperature index rather different from the NHM record (Figure 6). McIntyre and McKitrick [62] show that the 20th century was the warmest through the last 500 years but the 15th century was much warmer.

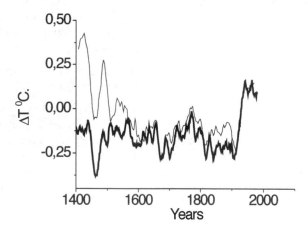

Figure 6. Tree-ring multi-proxy reconstructions of Mann et al. [54], and McIntyre and McKitrick [63], obtained using the same initial raw data. Both records are averaged over 13 years.

McIntyre and McKitrick [62, 63] stimulated the scientific community and generated some polemic discussion [41, 56, 64, 65, 92, 93]. The debate continues, but it can be most reasonably concluded that appreciable differences between the global temperature reconstructions obtained by means of the identical initial data indicate that, despite evident successes, the methods and approaches of paleoclimatology still leave considerable space for subjectivity. Taking into account the noted problems a question arises – what are the actual features of climate that could be captured by paleoreconstructions? That is to say: if the obtained temperature reconstructions have significant dissimilarities, what are their common features? Bürger [19] arrived at a conclusion that the available reconstructions differ so much that there is no way to draw meaningful conclusions from them. Ogurtsov et al. [73, 76] showed, however, that in spite of differences, the reconstructions of the Northern Hemisphere temperature have at least two apparent common features: (a) presence of a roughly regular century-scale rhythm with a period of 50–130 yrs through the last 1000 years; (b) a noticeable temperature rise during the last century. In spite of differences between IHS, MCV and MV reconstructions, they agree that the 20th century was warm. However, it is difficult to make any decisive conclusion about actual extent of the 20th century temperature anomaly within the last 1000 years, particularly if we take into account the possible influence of divergence or reduced sensitivity to temperature changes.

The *divergence problem* is a well known anomalous reduction in the sensitivity (ARS) of tree growth to changing temperature, which has been detected in many dendrochronological records over the last decades of the 20th century [9-10, 26, 31]. An evident underestimation of recent warming in tree-ring based reconstructions is illustrated in Figure 7.

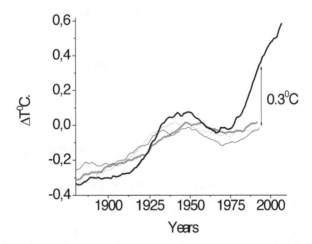

Figure 7. Thick black line – instrumentally measured temperature of extra-tropical Northern Hemisphere (http://data.giss.nasa.gov/gistemp/tabledata/ZonAnn.Ts+dSST.txt), thin black line – reconstruction of Jones et al. [43], dotted black line – reconstruction of Briffa [11], thick grey – reconstruction of Esper [30]. All the data sets smoothed over 15 years.

Some possible causes of the divergence are listed by D'Arrigo et al. [26], and [52]. They are:

a.   Decrease in stratospheric ozone concentration that causes increase of the flux of ultraviolet radiation reaching the ground and corresponding decline in tree productivity;

b.   Decrease of ground solar irradiance over 1961–1990 that caused decrease in sunlight reaching the ground;

c.   Nonlinear growth response, which causes reduction of tree-ring width under high temperatures.

In spite of some possible explanations the divergence problem still has not been solved and modern temperature reconstructions in dendrochronology usually do not cover well the most recent time interval. Therefore they may not capture well the sharp rise in temperature during the two last decades. Wilson et al. [95]) made a new temperature reconstruction for the Northern Hemisphere that utilizes fifteen tree-ring based proxy series that express no divergence effects over the last decades. Based on this divergence-free time series Ogurtsov et al. [76] corrected eight millennial-scale proxy reconstructions of temperature of the Northern Hemisphere for ARS effect and reanalyzed them. This study concluded that, neglecting the reconstruction [52], in the extratropical part of the Northern Hemisphere the time interval 1988–2008 was the warmest two decades in the last 1000 years with a probability of more than 0.70. The unusual level of current temperature over the areas least disturbed by local anthropogenic impact might prove that over the last two decades the climatic system was perturbed by an additional global-scale forcing factor, which did not operate in the past. It has been noted, however, that the procedure of correction for anomalous reduction in sensitivity includes some rather arbitrary assumptions and needs to be improved further (e.g. prolongation of linear trend over 2000–2008). [76].

Summarizing the results of analysis and reanalysis of the available long-scale temperature proxies, it could be confidently concluded that the 20th century was warm, i.e. that the global temperature averaged over the last 100 years was actually higher than global temperature averaged over the last 1000 years.

4. Modern climate modeling — Advantages and limitations

Climate models are mathematical descriptions of the Earth's climate system. They use sets of mathematical equations and numerical methods to reproduce the interactions of the atmosphere, mixed and deep ocean, land surface and cryosphere. Climate modeling has a variety of purposes from studies of the dynamics of the climate system to the prediction of future climate. Climate models range in complexity from simple, one-equation analytic models to state-of-the-art General Circulation Models (GCM), simulating the physics, chemistry, and biology of all the parts of the Earth's climatic system.

Energy-balance models of the globally averaged climate are the simplest. In the framework of the energy balance approach, changes in the climate system are estimated from an analysis of the change in the Earth's heat storage. The basis for these models was introduced by Budyko [15] and Sellers [85]. In its most simplified form, energy-balance model provides globally averaged values for the computed variables.

The more complicated radiative-convective models take into account the vertical variation of temperature with altitude. This approach makes it possible to study the role of clouds, water vapor and stratosphere. First radiative-convective model was introduced by Manabe and Wetherland [53].

GCMs are three-dimensional models with the boundary conditions at the spreading surface. They try to simulate incoming and outgoing radiation, time/spatial variation of the wind field, generation of clouds and transfer of water vapor, formation of sea ice, atmosphere-ocean coupling and redistribution of heat in oceans etc. GCMs have a spatial resolution comparable to the global synoptic network. The most prominent use of GCMs in recent years has been to forecast temperature changes resulting from increases in atmospheric concentrations of greenhouse gases [42]. GCMs are the most complex and sophisticated models including at least tens of equations.

It should be noted that while working with climatic models one needs to know a lot of model parameters, the number of which increases along with the increasing complexity of models. However, many such parameters are not known with adequate precision. This concerns even climatic sensitivity. The *climate sensitivity* (usually in $^0K \times W^{-1} \times m^2$) is a measure of the climate system's response to constant radiative forcing. *Radiative forcing* (usually in $W \times m^{-2}$) in turn is a measure of the perturbation brought by some factor in the radiative balance in the global Earth-atmosphere system. Positive forcing tends to warm the surface while negative forcing tends to cool it. Current estimations of the climate sensitivity, λ_c, range from 0.07 $^0K \times W^{-1} \times m^2$ [48] to 2.54 $^0K \times W^{-1} \times m^2$ [2]. IPCC [42] gives the value 0.53–1.23 $^0K \times W^{-1} \times m^2$ (see also Table 1 from [77]). Knowledge about radiative forcings which likely influenced terrestrial climate over the last 150 years – (a) anthropogenic greenhouse gas (CO_2, CH_4, N_2O) emissions; (b) anthropogenic aerosol emissions; (c) anthropogenic changes in albedo (land use, black soot on snow); (d) volcanic aerosol emissions; (d) total solar irradiance (TSI) variations – is still not satisfactory. According to [42] the level of scientific understanding is high only for industrial greenhouse gases. The level of understanding of all the other possible climate drivers is either medium or low. For example, forcing caused by human-made changes in land surface properties since pre-agricultural times is quite unclear. A possible value lies between 0.0 and 0.4 $W \times m^{-2}$ according to [42] and between 0.5 $W \times m^{-2}$ and –0.6 $W \times m^{-2}$ according to [72]. As a result an estimation of total net anthropogenic forcing since AD 1750 ranges from 0.6 $W \times m^{-2}$ to 2.4 $W \times m^{-2}$ [42]. Our knowledge about natural forcings is also rather limited. For example, different long-term reconstructions of TSI prior to instrumental period (before 1978) show a quite different picture. According to [87], the average TSI increased by ca 5 $W \times m^{-2}$ from the beginning of the 19th century till the end of the 20th century, while the reconstruction [68] shows only 1.5 $W \times m^{-2}$ rise during the same time interval (see also Figure 6 from [77]). These values lead to a corresponding radiative forcing 0.26–0.88 $W \times m^{-2}$. Furthermore, considerable uncertainty remains over the magnitude of influence of volcanic eruptions on the climate system. Even in the case of Mt. Pinatubo explosion (1991), which was directly observed and investigated using all the possibilities of modern science, the estimations of its climatic forcing differ from 2.25 W×m⁻² [71] to 4.7 W×m⁻² [1]. Knowledge about the structure of feedbacks is also incomplete [46]. Thus, it is obvious that the parameters of many modern climate models have huge uncertainties. That is why some skeptics even believe that they follow the old maxim of "garbage in, garbage out" – the principle in the computer science, meaning that if the input data are incorrect then erroneous results would be obtained even if the algorithms are correct. It is interesting to note that a wide spread in the main input data and model parameters actually makes it possible to fit the calculated temperature to the measured one in rather different ways. Numerical experiment of [74] shows that an arbitrary choice of radiative forcings without justification of the choice criteria makes it possible to explain to a great extent the global warming of the 20th century beyond the hypothesis about the greenhouse effect. Ogurtsov [73] tested the total (direct and indirect) contribution of the solar activity to the global warming using one-dimensional energy-balance climate model. This work takes into account the fact that the Sun can affect the Earth's climate not only directly, via changes in solar luminosity, but probably also indirectly via the modulation of galactic cosmic ray (GCR) flux. A correlation between the changes in the globally averaged low (<3.2 km in altitude) cloud cover anomaly and the changes in the GCR intensity was indeed demonstrated by Marsh and Svensmark [58, 59] and Palle et al. [79]. Data on low cloudiness obtained in the framework of the International Satellite Cloud Climate Project (ISCCP) are plotted in Figure 8 together with the data on GCR flux measured by neutron monitor in Kiel.

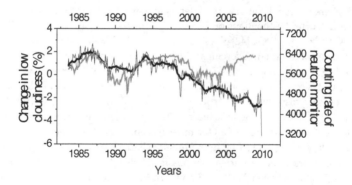

Figure 8. Black thin line – monthly data on the global average of low (>680 hPa) cloud cover anomalies (ftp://isccp.giss.nasa.gov/pub/data/D2CLOUDTYPES); thick black line – yearly averages. Grey line – counting rate of the Kiel neutron monitor (http://cr0.izmiran.rssi.ru/kiel/main.html).

One can see appreciable positive correlation between GCR and low cloudiness through 1983–2001. Ogurtsov [74] calculated the hypothetical cloud radiative forcing since the end of the 19th century based on: (a) linear relationship between low cloudiness and GCR over 1983–1994, proposed by Marsh and Svensmark [58]; (b) long-term reconstruction of GCR intensity obtained by Mursula et al. [70]; (c) estimations of cloud radiative forcing made using the data of the Earth Radiation Budget Experiment [58]. Using this forcing, TSI reconstruction after Hoyt and Schatten [39] (Figure 9B) and a simple one-dimensional (4 latitudinal belts) energy-balance model, Ogurtsov [74] calculated the mean temperature in the Northern Hemisphere over 1886–1999 (Figure 9C). It is evident that the joint effect of the changes in: (a) the solar luminosity and (b) low cloudiness may lead to an increase in the hemispheric temperature in the 20th century by about 0.35°C. Thus, in the framework of the model of [74], the warming of the Northern Hemisphere before the 1980s can be fully accounted for changes in solar-cosmic factors with the greenhouse effect fully neglected.

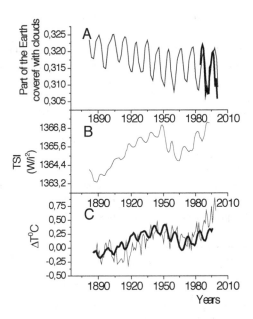

Figure 9. A – Low cloudiness over the middle latitudes (25–60° N) of the Northern Hemisphere. Thick line – experimental data (yearly averages), thin line – reconstruction from the data on GCR intensity obtained by Mursula et al. [70]. B – reconstruction of the solar luminosity after Hoyt and Schatten [39]. C – mean annual temperature of the Northern Hemisphere. Thin line – instrumental data, thick line – model calculation [74].

However, a decrease of the correlation between GCR and clouds at the end of the 1990s and a change of its sign after 2002 makes any conclusion about possible cosmic ray-cloud link rather disputable [29, 88]. The calculation of [74] is thus only a computing exercise, which shows that an arbitrary manipulation of the input model parameters, neither of which is known precisely and reliably enough, can bring us quite curious results. This result as well as many other points (see e.g. Lindzen [49], testify that despite large successes in climate modeling, current climate simulations seem still to be only fitting of the calculation results to the actual observed temperatures. A fairly good fitting can be achieved using various combinations of input data. Such investigations are very important since they are necessary for studying possible scenarios of climatic changes in the past as well as in the future. However, at present it is not easy to estimate (even approximately) the probability of each specific scenario. We can only note that solar contribution to the sharp temperature increase during the last 3–4 decades is either minor [89] or negligible [7]. The rise of temperature after 1980 has not been simulated by the model of [74] (see Figure 9C). This testifies that the observed rapid rise in global mean temperatures after the beginning of 1970s is difficult to reproduce in any model if the greenhouse gas forcing is not taken into account.

5. Possible scenarios of climate change in the 21st century

Projection of forthcoming climatic changes is one of the main tasks of climatology. There are two approaches to understanding future evolution of climate: (a) mathematical and (b) physical. In the mathematical approach we take the observed climate record and try to extrapolate it properly into the future. In the physical approach we first attempt to understand the most important climate processes and describe them with a detailed model. Then the obtained model is used to predict the response of future climate to different forcings. Attempts at predicting future warming of the globe started in the 1970s–1980s. These physical forecasts were made assuming a predominantly greenhouse character of GW and were based on prognoses of future CO_2 concentrations. The concentration of carbon dioxide has been predicted quite correctly, e.g. Legasov et al. [47] forecasted a 375–385 ppm value of CO_2 in the year 2000 while Bolin et al. [8] estimated a corresponding value of 360–380 ppm. However, the predictions of global temperature were not equally successful (Table 2).

Source	The forecast formula	Predicted temperature in 2000 (T °C)
Budyko, 1972 [16]	$T_{2000}=T_{1900}+1.2$	0.83–1.00
Kellogg , 1978 [45]	$T_{2000}=T_{1900}+1.2$	0.83–1.00
Budyko, 1982 [17]	$T_{2000}=\bar{T}(1880-1975)+0.6$	0.35–0.49
The impact of atmospheric carbon dioxide increasing on climate, 1982 [94]	$T_{2000}=T_{1900}+(1.0-2.0)$	0.63–1.80
Budyko and Izrael 1991, [18]	$T_{2000}=T_{1970}+0.9$	0.72–0.91

Table 2. Forecasts of a mean global temperature in the year 2000 made during 1972–1987. Actual temperature in 2000 was 0.27–0.39 T °C.

Real temperatures in the years 1900, 1970 and 2000 were determined by means of the data of CRU (http://www.cru.uea.ac.uk/cru/info/warming/) and NCDC (ftp://ftp.ncdc.noaa.gov/pub/

data/ anomalies/monthly.land_ocean.90S.90N.df_1901-2000mean.dat). The majority of the predictions (Table 1) clearly overestimated the warming at the end of the 20th century, since the real instrumental temperature in 2000 reached 0.27 (CRU) and 0.39 (NCDC). This is also true for the detailed prognosis made by a group headed by Hansen et al. [35]. These researchers considered three possible scenarios based on rising concentrations of greenhouse gases (CO_2, CH_4, N_2O, CFC_{11}, CFC_{12}) till 2020. Scenario A assumes continued exponential increase in greenhouse gases, scenario B assumes a reduced linear growth and scenario C assumes a rapid reduction of emissions such that the net climate forcing ceases to increase after the year 2000 (see Fig. 10A). Hansen et al. [35] have calculated respective variations in global temperature for these scenarios (Figure 10B). Even the minimum scenario of [35] considerably (up to 0.1℃) overestimates the value of the actual measured temperature during 2005–2010. Naturally our knowledge on a climatic system has appreciably increased since the end of the 1980s. Nevertheless, the evident failure of the early GW predictions seems rather indicative. It proves that more or less reliable prediction of the future climate evolution is a very complicated task. Insufficient knowledge about possible and potential forcing factors deteriorates the reliability of climatic modeling and, hence, reduces opportunities of the physical methods of forecast. In addition, discrepant knowledge about the history of climate prevents us from more or less reliable extrapolation of temperature into the future. Consequently these shortcomings limit the applicability of the mathematical methods. As a conclusion based on the available climatic and paleoclimatic data two possible scenarios of the global temperature change in the 21st century are considered. These scenarios in turn are based on extreme scenarios of the climate evolution during the last 100 years.

a. GW is unique and unprecedented. It is connected mainly with an extra global-scale forcing factor, which did not operate in the past. It is clear that anthropogenic greenhouse effect is the main additional contributor and the temperature history of last 1000 years, which is compatible with the IHS reconstructions. Global warming appears almost entirely as a result of industrial activity of mankind, and the past of the climatic system does not play appreciable part (neglecting the natural factors). These projections, which are based on various assumptions on the future evolution of industrial emissions of greenhouse gases, result in temperature rise of the 21st century between 1.8–4.0℃ [42].

b. GW is not unique in a historical context and it is mainly the result of natural (terrestrial, solar, cosmophysical) climatic cycles while contribution of greenhouse effect is of minor importance. In that case the temperature history of the last 1000 years is compatible with MV and MCV reconstructions. Climate in the 20th century was driven by the same dynamic system as during the entire last millennium (neglecting the greenhouse effect). This in turn means that the present state of climate is a natural result of its past, and the paleoclimatic data (MV and MCV proxies) could be used as a source of information for forecasts. For this purpose we interpolated the paleorecords NHL and NHMb by decades, and made a prognosis of mean decadal temperature over the first part of the 21st century by means of a nonlinear forecast technique. The nonlinear prediction was made using the method of analogs, which is based on the reconstruction of the trajectory of the predicted series dynamic system in a pseudo-phase space and is a modification of the method used in [90].

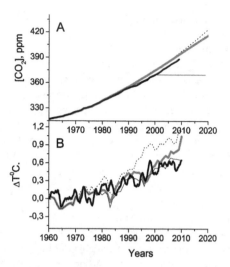

Figure 10. A – Concentration of CO_2 in the Earth's atmosphere: black thick line – experimental measurement at Mauna-Loa (www.esrl.noaa.gov/gmd/ccgg/trends/), dotted line – scenario A by Hansen et al. [35], gray line – scenario B [35], black thin line – scenario C [28]. B – Global surface temperature computed for scenarios A, B, and C, compared with observational data: black thick line – instrumental measurement (ftp://ftp.ncdc.noaa.gov/pub/data/anomalies/monthly.land_ocean.90S.90N.df_1901-2000mean.dat), other lines – corresponding scenarios [35].

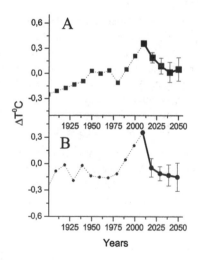

Figure 11. Forecasts of future Northern Hemisphere temperatures. A – Prediction based on reconstruction by Loehle [51]. B – Prediction based on the reconstruction by Moberg et al. [67]. Both proxy series were extrapolated till 2010 by means of the instrumental temperature data. The original data, averaged by 13 years and interpolated by decades, are shown with dotted lines. Predicted values are shown with thick black lines. This prediction was made using the embedding dimension d = 3 and seven nearest neighbors.

If the warming of the last 100 years is a result of natural climatic variability, i.e. the climate of past century is governed by the same dynamic system as the previous one to two millennia, the mean temperature of the Northern Hemisphere in the first part of the 21st century will unlikely be higher than the modern value Figure 11. Ogurtsov and Lindholm [72] forecasted the same.

If the GW is a result of a variety of different processes of both natural and anthropogenic origin, neither of which could be neglected (greenhouse gas emission, solar activity change, natural climatic variation, regional and local anthropogenic impact), the situation is most complicated. In that case it is challenging to make even a qualitative estimation of changes of a global climate through current century because of the significant uncertainty in our knowledge on the relative contributions of the specified factors.

6. Conclusion and prospects for further research

The analysis of the available information on the temperature of the Earth, including both instrumental temperature measurements and proxy paleodata, leads to the conclusion that the 20th century was warm, i.e. the average temperature of the Earth through AD 1900–2000 was undoubtedly higher than the average temperature through AD 1000–2000. The forcing factors, which presumably cause the global warming are: (a) anthropogenic changes in the atmospheric concentration of greenhouse gases and aerosols, (b) changes in solar activity, (c) internal oscillations in the climatic system, (d) changes in volcanic activity, and (e) anthropogenic changes in land surface properties. Greenhouse forcing is often reported as the major contributor to GW. Climatic modeling gives evidence in favor of this assumption, since it is very hard to simulate abrupt rise of temperature starting in 1970s, if greenhouse gas influence is neglected. Nevertheless greenhouse skeptics propose a few ideas to explain the phenomenon. Pokrovski [82] suggests that the sharp recent warming is mainly a result of a strong 60–70 years natural temperature cycle connected with the circulation of oceanic water. The hypothesis sounds plausible but it should be noted that the interval of instrumental measurement is not long enough to establish both the phase and amplitude of this cycle accurately. Analyses of paleodata confirms the presence of the century-scale cyclicity in Northern Hemisphere temperature [73] but its peak-to-trough amplitude unlikely exceeds 0.3 °C and the second part of 20th century seems to be a decline phase of this variation (see Fig.4 of [73]). Another idea was considered by Bashkirtsev and Mashnich [4] who suggest taking into consideration the globally averaged satellite cloud observations made in the framework of ISCCP. Bashkirtsev and Mashnich [4] propose that the change in the Earth's albedo, caused by downward trend in multi-decadal record of cloudiness, is sufficient to provide the observed temperature variations at the end of the 20th century. Their oversimplified estimation showed that a decrease in a global cloud area during 1987–2000, which usually is not considered by climatic models, could cause increase in background solar radiation flux up to 10 W m^{-2}. Moreover, Palle et al. [80, 81] claimed that a change in the Earth's reflectance, tightly related to the cloudiness, has resulted in appreciable increase in solar radiation incident at the Earth's surface at the end of the 20th century. The phenomenon is often called a global brightening. The more detailed estimations of [81], based on both the data of satellite and ground-based astronomical and actinometrical observations, show that the respective radiative forcing reaches 2–7 W × m^{-2} during 1985–2000. Calculations of [78] show that despite a short period of action this forcing factor should result in a corresponding very sharp rise of a global temperature. It should be noted, however, that the quality of the ISCCP cloud data is doubtful [25] and particularly long-term trends of cloudiness, established by means of satellite measurement, are highly disputable [32]. That is why the question about changes in the Earth' albedo during the last decades is still open.

The possible failure of the models to predict the troposphere warming, particularly in the tropics, seems to be the most serious challenge to greenhouse theory now. Disagreement between model prediction and observational data is repeatedly used by greenhouse skeptics as evidence about the poor quality of climate models. Discussions created by this controversy are often beyond purely scientific debates and concern some philosophical issues, e.g.: if model prediction disagrees with the experimental results, which one is most likely wrong – model or experiment [50]? Solutions to the problems concern the tropical troposphere warming, which thus is a crucial point for understanding the origin of the GW.

Paleoclimatic data unlikely improve our understanding of the CO_2-temperature change relationship appreciably. Indeed, research of Antarctic ice-core paleorecords, covering the last 9–420 kA, reveal a pattern of strong temperature and CO_2 rises at roughly 100 000-year intervals. But during these great temperature transitions the CO_2 rise has almost always come 400–5000 years after (not before) the temperature increase [20, 69]. This link most likely appears as warmer temperatures have facilitated release of the gas from oceans. Therefore ice-core paleodata give evidence that temperature controls concentration of carbon dioxide in the atmosphere and not vice versa. Greenhouse warming supporters do not deny this conclusion but emphasize that this cause-effect relationship took place in the past while in the case of a contemporary warming, the external climate forcing by anthropogenic CO_2 emissions leads climate variations. Thus the application of the CO_2-climate relation deduced from the past on a recent global warming is not fully substantiated [33]. Moreover a study by Shakun et al. [86], who examined 80 proxy records from around the globe 20–10 kyr ago (the last glacial-interglacial transition), showed that the temperature rise happened first in the Southern Hemisphere, while in the Northern Hemisphere the CO_2 increase was first. Shakun et al. [86] arrived at conclusion that about 90% of the global warming occurred after the CO_2 increase.

Based on the totality of the available paleodata we can infer that global temperature during the last 2–3 decades was:

a.   certainly highest over the last 500 years.

b.   probably highest over the last 1000 years. However, it is not possible to conclude reliably that the last 20–30 years was the warmest period of the entire millennium, because the existing reconstructions of temperature during the last 600–1000 years depict different and even discrepant patterns. This disagreement can hardly be explained by uncertainties inherent to the proxies. It should also be noted that neither of the temperature histories based on the paleodata can be considered satisfactorily precise and reliable, due to e.g. the insufficient coverage of the Earth's surface by individual paleorecords.

Moreover, it is difficult to estimate the contribution of any individual factors potentially responsible for the GW – industrial emission of greenhouse gases, varying activity of the Sun, regional anthropogenic impact and natural climatic cycles – due to insufficient knowledge of the corresponding radiative forcings and climate sensitivity. Actually, the estimation of total net anthropogenic forcing since 1750, made by IPCC [42], gives a value 0.6–2.4 W×m^{-2}. Our estimation of direct solar forcing, caused by change in luminosity since the beginning of the 19th century, gives a value 0.26–0.88 W×m^{-2}. If we use the assessment of climate sensitivity [42] $\lambda_c = 0.53$–1.23 ^{0}K×W^{-1}×m^2 we obtain a corresponding warming by 0.32–2.95^{0}C due to anthropogenic factor and by 0.14–1.08^{0}C due to change in TSI. The difference between the lower and upper limits reaches almost an order of magnitude. In addition, there are other studies indicating that the Sun can affect terrestrial climate indirectly, e.g.: (a) via a connection between the cloud cover and galactic cosmic ray intensity [58, 59] and (b) via a connection between the galactic and solar cosmic ray intensity and aerosol content [75]. However, these hypotheses have not been reliably proven and thus it is impossible to obtain quantitative estimation of corresponding forcings.

Temperature reconstructions of MV and MCV types show that natural cycles of longer scale with larger amplitudes can also be an important factor of the GW. But it is very difficult to determine the actual role of internal variability of the climatic system in the GW because of disagreement between different proxies and their limited precision and reliability.

Summarizing all stated above, we can conclude that the origin of the rise of global temperature should be considered as not well known due to a lack of adequate knowledge about many of the factors that may be responsible for this phenomenon. Consequently, it is very difficult to predict the climatic change in the 21st century even nearly precisely. The available information allows only specifying two possible scenarios of the evolution of global temperature during this century:

a. If global warming is almost entirely a result of industrial greenhouse effect the average temperature of the globe in the 21st century will continue to increase significantly, in agreement with the projections of IPCC.

b. If global warming is only slightly connected with the anthropogenic activity and is primarily a result of natural climatic variability (a less probable scenario), then the average temperature of the Northern Hemisphere will not increase at least during the first half of the 21st century.

Undoubtedly many other climatic scenarios are possible as well. However, it seems unlikely that the problem of the origin of the modern increase of global temperature will be solved before we reach the years in the middle of the current century. Substantial improvement of both climate modeling and experimental monitoring of the current state of the atmosphere is of great importance to establish origin and character of the GW definitely. Further progress in paleoclimatology can also help to solve the problem.

Acknowledgements

M. G. Ogurtsov expresses his thanks to the exchange program between the Russian and Finnish Academies (project No. 16), to the program of the Presidium of RAS No 22, and to RFBR grants 10-05-00129, 11-02-00755 for financial support. R. Jalkanen and M. Lindholm acknowledge The Finnish Academy (grant SA 138 937). Authors are thankful to the Editor for the constructive comments which greatly helped to improve the chapter.

Author details

M.G. Ogurtsov[1*], M. Lindholm[2] and R. Jalkanen[2]

*Address all correspondence to: maxim.ogurtsov@mail.ioffe.ru

1 Ioffe PhTI, St. Petersburg, Russia, Central Astronomical Observatory at Pulkovo, S. Petersburg, Russia

2 Finnish Forest Research Institute, Rovaniemi, Finland

References

[1] Andronova NG, Rozanov EV, Yang F, Schlesinger ME, Stenchikov G L. Radiative Forcing by Volcanic Aerosols from 1850 to 1994. Journal of Geophysical Research 1999;104 16 807–16 821.

[2] Andronova NG, Schlesinger ME. Causes of Global Temperature changes during the 19th and 20th centuries. Geophysical Research Letters 2000; 27(14) 2137-2140.

[3] Barnett TP, Santer BD, Jones PD. Estimates of Low Frequency Natural Variability in Near-Surface Air Temperature. The Holocene1996;6.3 225-263.

[4] Bashkirtsev VS, Mashnich GP. The Sun and the Earth's climate. In: Proceedings of the Russian annual conference on the solar physics: Astronomy year: solar and solar-terrestrial physics: St.Petersburg; 2009, p55-58 (in Russian).

[5] Basurah HM. Nile Flooding fluctuations and its possible connection to the long solar variability. Journal of the Association of Arab Universities for Basic and Applied Sciences 2005;1 1-7

[6] Beltrami H. Climate from Borehole Data: Energy Fluxes and Temperatures Since 1500. Geophysical Research Letters 2002;29(23) 2111, doi:10.1029/2002GL015702.

[7]   Benestad R E, Schmidt GA. Solar Trends and Global Warming. Journal of Geophysi-
      cal Research 2009; 114, D14101, doi:10.1029/2008JD011639

[8]   Bolin B, Doos BR, Jager J, Warrick RA. The Greenhouse Effect, Climate Change and
      Ecosystems. SCOPE 1986; 29. John Wiley & Sons, New York.

[9]   Briffa K, Schweingruber F, Jones P, Osborn T. Reduced Sensitivity of Recent Tree
      Growth to Temperature at High Northern Latitudes. Nature 1998a;391 678-682.

[10] Briffa K, Schweingruber F, Jones P, Osborn T, Harris I, Shiyatov S, Vaganov A,
      Grudd H. Trees Tell of Past Climates: But are They Speaking Less Clearly Today?
      Philosophical Transactions of the Royal Society B 1998b;353 65-73.

[11] Briffa KR. Annual Climate Variability in the Holocene: Interpreting the Message of
      Ancient Trees. Quaternary Science Reviews 2000;19 87-105.

[12] Briffa KR, Osborn TJ, Schweingruber FH, Harris IC, Jones PD, Shiyatov SG, Vaganov
      EA. Low frequency temperature variations from a northern tree-ring density net-
      work. Journal of Geophysical Research 2001;106 2929–2941.

[13] Briffa KR, Osborn TJ. Blowing Hot and Cold. Science 2002;295 2227-2228.

[14] Brohan P, Kennedy JJ, Haris I, Tett SFB, Jones PD. Uncertainty Estimates in Regional
      and Global Observed Temperature Changes: a New Dataset From 1850. Journal of
      Geophysical Research 2006;111 D12106, doi:10.1029/2005JD006548,

[15] Budyko MI. The Effect of Solar Radiation Variations on the Climate of the Earth. Tel-
      lus 1969;21(5) 611-619.

[16] Budyko MI. Man's Influence on Climate. Leningrad: Gidrometeoizdat; 1972 (in Rus-
      sian).

[17] Budyko MI. The Earth's climate: past and future. Academy Press: New York; 1982.

[18] Budyko MI, Izrael YA. Anthropogenic Climate Change. University of Arizona Press:
      Tucson; 1991.

[19] Bürger G. Clustering climate reconstructions. Climate of the Past Discussions 2010;6
      659-679.

[20] Caillon N, Severinghaus JP, Jouzel J, Barnola, J-M, Kang J, Lipenkov VY. Timing of
      Atmospheric CO_2 and Antarctic Temperature Changes Across Termination III. Sci-
      ence 2003;299 (5613) 1728-1731.

[21] Christy JR, Norris WB What may we conclude about global tropospheric tempera-
      ture   trends?   Geophysical   Research   Letters   2004;31   L06621,   doi:10.1029/
      2003GL019361.

[22] Christy JR, Spencer RW, Mears CA, Wentz F. Correcting Temperature Data Sets. Sci-
      ence 2005;310(5750) 972-973.

[23] Cook ER, Kairiukstis LA Methods of Dendrochronology: Applications in the Environmental Sciences. Kluwer Academic Publishers: Dordrecht; 1989.

[24] Crowley TJ, Lowery TS. How Warm Was the Medieval Warm Period? Ambio 2000;29 51-54.

[25] Dai A, Karl TR, Sun B, Trenberth KE. Recent Trends in Cloudiness over the United States. A tale of monitoring inadequacies. Bulletin of American Meteorological Society 2006; 597-606, DOI:10.1175/BAMS-87-5-597

[26] D'Arrigo R, Wilson R, Liepert B, Cherubini P. On the 'Divergence Problem' in Northern Forests: A Review of The Tree-Ring Evidence and Possible Causes. Global and Planetary Change 2008;60 289–305.

[27] Dean JS, Meko DM, Swetman TW Tree Rings, Environment and Humanity. Radiocarbon. Tuscon: University of Arizona press; 1996.

[28] Douglass DH, Christy JR, Pearson BD, Singer SF. A Comparison of Tropical Temperature Trends With Model Predictions. International Journal of Climatology 2008;28 1693–1701.

[29] Erlykin AD, Wolfendale AW. Cosmic Ray Effects on Cloud Cover and Their Relevance to Climate Change. Journal of Atmospheric and Solar-Terrestrial Physics 2011;73(13) 1681-1686.

[30] Esper J, Cook ER, Schweingruber F.H. Low-Frequency Signals in Long Tree-ring Chronologies for Reconstructing Past Temperature Variability. Science 2002;295(5563) 2250-2253.

[31] Esper J, Frank D. Divergence Pitfalls in Tree-Ring Research. Climatic Change 2009;94 261–266.

[32] Evan AT, Heidinger AK, Vimont DJ. Arguments Against a Physical Long-Term Trend in Global ISCCP Cloud Amounts. Geophysical Research Letters 2007:34, L04701, doi:10.1029/2006GL028083.

[33] Fischer H, Wahlen M, Smith J, Mastroianni D, Deck B. Ice Core Records of Atmospheric CO_2 Around the Last Three Glacial Terminations. Science 1999;283 (5408) 1712-1714.

[34] Fritts H. Tree rings and climate. Academic Press: London; 1976.

[35] Hansen J, Fung I, Lacis A, Rind D, Lebedeff S, Ruedy R, Russell G, Stone P. Global Climate Changes as Forecast by Goddard Institute for Space Studies Three-Dimensional Model. Journal of Geophysical Research 1988;93(D8) 9341-9364.

[36] Fu Q, Manabe S, Johanson CM., On the Warming in the Tropical Upper Troposphere: Models Versus Observations. Geophysical Research Letters 2011;38 L15704, doi: 10.1029/2011GL048101.

[37] Helama S, Timonen M, Lindholm M, Merilälnen J, Eronen M. Extracting Long-Period Climate Fluctuations from Tree-Ring Chronologies Over Timescales of Centuries to Millennia. International Journal of Climatology 2005;25 1767-1779.

[38] Heymann M. The Evolution of Climate Ideas and Knowledge. WIREs Climate Change 2010;1 581–597.

[39] Hoyt DV, Schatten KH. A discussion on plausible solar irradiance variations, 1700-1992. Journal of Geophysical Research 1993:98(A11) 18895-18906.

[40] Huang SP, Pollack HN, Shen PY. Late Quaternary Climate Reconstruction Based on Borehole Heat Flux Data, Borehole Temperature Data and the Instrumental Record. Geophysical Research Letters 2008;(35) L13703.

[41] Huybers P. Comment on "Hockey Sticks, Principal Components, and Spurious Significance" by S. McIntyre and R. McKitrick. Geophysical Research Letters 2005;32 L20705, doi:10.1029/2005GL023395.

[42] IPCC. WG1 Fourth Assessment Report: Climate Change 2007: The Physical Science Basis: Summary for Policymakers. Paris; 2007.

[43] Jones PD, Briffa KR, Barnett TP, Tett SFB. High-Resolution Palaeoclimatic Records for the Last Millennium: Interpretation, Integration and Comparison with General Circulation Model Control-Run Temperatures. The Holocene 1998;8.4 455-471.

[44] Jones PD, Parker DE, Osborn TJ, Briffa KR.. Global and hemispheric temperature anomalies – land and marine records. In Trends: A compendium of data on global change. Carbon dioxide information analysis center, Oak Ridge National Laboratory, US Department of Energy: Oak Ridge, Tennessee, USA; 2001

[45] Kellogg W. Review of mankind's impact on global climate. In: Multidisciplinary research related to the atmospheric sciences: Boulder; 1978, p64-81

[46] Knutti R, Allen MR., Friedlingstein P, Gregory JM, Hegerl GC, Meehl GA, Meinshausen M, Murphy JM, Plattner G-K, Raper SCB, Stocker TF, Stott PA, Teng H, Wigley TML. A Review of Uncertainties in Global Temperature Projections over the Twenty-First Century. Journal of Climate 2008;(21) 2651-2663.

[47] Legasov VA; Kuzmin II, Chernoplekov AN. Izvestiya AN SSSR, Atmospheric and Oceanic Physics 1984;20(11) 1089-1106 (in Russian).

[48] Lindzen RS, Giannitsis C. On the Climatic Implications of Volcanic Cooling. Journal of Geophysical Research 1998;103(D6) 5929-5941.

[49] Lindzen P. http://phys.org/news/2012-07-climate-flawed-speaker-sandia.html; 2012

[50] Lloyd EA. The Role of 'Complex' Empiricism in the Debates About Satellite Data and Climate Models. Studies in History and Philosophy of Science 2012;43 390–401.

[51] Loehle CA 2000-year global temperature reconstruction based on non-treering prox-ies. Energy and Environment 2007;18(7-8) 1049-1058.

[52] Loehle C. A Mathematical Analysis of the Divergence Problem in Dendroclimatolo-gy. Climatic Change 2009;94 233–245.

[53] Manabe S, Wetherland RT Thermal Equilibrium of the Atmosphere with a Given Distribution of Relative Humidity. Journal of the Atmospheric Science 1967;24(3) 241-259.

[54] Mann ME, Bradley RS, Hughes MK. Northern Hemisphere Temperatures During the Past Millennium: Inferences, Uncertainties, and Limitations. Geophysical Research Letters 1999;26(6) 759-762.

[55] Mann M, Hughes M. Tree Ring Chronologies and Climate Variability. Science 2002;296(5569) 848-852.

[56] Mann ME, Bradley RS, Hughes MK. Note on Paper by McIntyre and McKitrick in: Energy and Environment, ftp://holocene.evsc.virginia.edu/pub/ mann/EandEPaper-Problem.pdf; 2003

[57] Mann MN, Zhang Z, Hughes MK, Bradley RS, Miller SK, Rutherford S, Fenbiao N. Proxy-Based Reconstructions of Hemispheric and Global Surface Temperature Varia-tions over the Past Two Millennia. Proceedings of the National Academy of Sciences 2008;105(36) 13252-13257.

[58] Marsh N, Svensmark H. Low Cloud Properties Influenced by Cosmic Rays. Physical Review Letters 2000,85(23) 5004-5007.

[59] Marsh N, Svensmark H. Galactic Cosmic Ray and El Nin˜o-Southern Oscillation Trends in ISCCP-D2 Low-Cloud Properties. Journal of Geophysical Research 2003;108(D6) 4195-4199.

[60] McCarroll D, Jalkanen R, Hicks S, Tuovinen M, Gagen M, Pawellek F, Eckstein D, Schmitt U, Autio J, Heikkinen O. Multiproxy Dendroclimatology: a Pilot Study in Northern Finland. The Holocene , 2003;13(6) 829–838.

[61] McCarroll D, Loader N, Jalkanen R, Gagen M, Grudd H, Gunnarson B, Kirchhefer A, Kononov Y, Boettger T, Friedrich M, Linderholm H, Lindholm M, Los S, Remmele S, Yamazaki H, Young G, Zorita E. A 1200-year Multi-Proxy Record of Tree Growth and Summer Temperature at the Northern Pine Forest Limit of Europe. The Holo-cene 2012 (in press),

[62] McIntyre S, McKitrick R. Corrections to the Mann et al (1998) Proxy Data Base and Northern Hemisphere Average Temperature Series. Energy and Environment 2003;14(6) 751-772.

[63] McIntyre S, McKitrick R. Hockey Sticks, Principal Components and Spurious Signifi-cance. Geophysical Research Letters 2005a;32 L03710, doi:2004GL021750.

[64] McIntyre S, McKitrick R. Reply to Comment by von Storch and Zorita on "Hockey Sticks, Principal Components and Spurious Significance". Geophysical Research Letters 2005b;32 L20714, doi:10.1029/2005GL023089.

[65] McIntyre S, McKitrick R. Reply to Comment by Huybers on "Hockey Sticks, Principal Components and Spurious Significance", Geophysical Research Letters 2005c;32 L20714, doi:10.1029/2005GL023586,

[66] Mears CA, Wentz FJ. The Effect of Diurnal Correction on Satellite-Derived Lower Troposphere Temperature. Science 2005;309(5740) 1548-1551.

[67] Moberg A, Sonechkin DM, Holmgren K, Datsenko MM, Karlen W. High Variable Northern Hemisphere Temperatures Reconstructed from Low- and High-Resolution Proxy Data. Nature 2005;433(7026) 613-617.

[68] Mordvinov AV, Makarenko NG, Ogurtsov MG, Jungner H. Reconstruction of Magnetic Activity of the Sun and Changes in its Irradiance on a Millennium Timescale Using Neurocomputing. Solar Physics 2004;224 247-253.

[69] Mudelsee M. The Phase Relations Among Atmospheric CO_2 Content, Temperature and Global Ice Volume Over the Past 420 ka. Quaternary Science Reviews 2001;20 583-589.

[70] Mursula K, Usoskin IG, Kovaltsov GA. Reconstructing the Long-Term Cosmic Ray Intensity: Linear Relations Do Not Work. Annales Geophysicae 2003;21 863-867.

[71] Myhre G, Myhre A, Stordal F. Historical Evolution of Radiative Forcing of Climate. Atmospheric. Environment 2001;(35) 2361–2373.

[72] Ogurtsov MG, Lindholm M. Uncertainties in Assessing Global Warming During the 20th Century: Disagreement Between Key Data Sources. Energy and Environment 2006;17(5) 685-706.

[73] Ogurtsov MG, Jungner H, Lindholm. A Potential Century-Scale Rhythm in Six Major Paleoclimatic Records in the Northern Hemisphere. Geografiska Annaler 2007;89A(2) 129-136.

[74] Ogurtsov MG. On the Possible Contribution of Solar-Cosmic Factors to a Global Warming of 20th Century. Izvestiya RAN, Physics 2007A;71(7) 1051-1053.

[75] Ogurtsov MG. Secular Variation in Aerosol Transparency of the Atmosphere as the Possible Link Between Long-Term Variations in Solar Activity and Climate. Geomagnetism and Aeronomy 2007B;47(1) 118-128.

[76] Ogurtsov MG, Jungner H, Helama S, Lindholm M, Oinonen M. Paleoclimatological Evidence for Unprecedented Recent Temperature Rise at the Extratropical Part of the Northern Hemisphere. Geografiska Annaler 2011;93(1) 17-27.

[77] Ogurtsov M, Lindholm M, Jalkanen R. Solar Activity, Space Weather and the Earth's Climate, In: Hannachi A. (ed.) Climate variability - some aspects, challenges and prospects. Rieka: InTech;2012. p39-72.

[78] Ogurtsov M, Lindholm M, Jalkanen R. Background Solar Irradiance and the Climate of the Earth in the End of the 20th Century. Atmospheric and Climate Sciences 2012;2(2) 191-195.

[79] Palle E, Buttler CJ, O'Brien K. The Possible Connection Between Ionization in the Atmosphere by Cosmic Rays and Low Level Clouds. Journal of Atmospheric and Solar-Terrestrial Physics 2004; 66(2) 1779-1790.

[80] Palle E, Montanes-Rodrigues P, Goode PR, Koonin SE, Wild M, Casadio S. A Multi-Data Comparison of Shortwave Climate Forcing Changes. Geophysical Research Letters 2005;32 L21702.

[81] Palle E, Goode PR, Montanes-Rodriguez P, Koonin SE. Can Earth's Albedo and Surface Temperatures Increase Together? EOS 2006;70 (4) 37-43.

[82] Pokrovsky O.M. The Analysis of Factors of Climate Change According to Remote and Contact Measurements. Research of the Earth from Space 2010;5 11-24 (in Russian).

[83] Rutherford S, Mann ME. Correction to "Optimal surface temperature reconstructions using terrestrial borehole data". Journal of Geophysical Research 2004;109 D11107, doi:10.1029/2003JD004290.

[84] Santer BD, Thorne PW, Haimberger L, Taylor KE, Wigley TML, Lanzante JR. Consistency of Modeled and Observed Temperature Trends in the Tropical Troposphere. International Journal of Climatology 2008;28 1703-1722.

[85] Sellers WD. A Global Climatic Model Based on the Energy Balance of the Earth-Atmosphere System. Journal of Applied Meteorology 1969;8(3) 392-400.

[86] Shakun JD, Clark PU, He F, Marcott SA, Mix AC, Liu Z, Otto-Bliesner B, Schmittner A, Bard E. Global Warming Preceded by Increasing Carbon Dioxide Concentrations During the Last Deglaciation. Nature 2012;484 (7392) 49-54.

[87] Shapiro AI, Schmutz W, Rozanov E. A New Approach to the Long-Term Reconstruction of The Solar Irradiance Leads to Large Historical Solar Forcing. Astronomy and Astrophysics 2011;529 A67.

[88] Sloan T, Wolfendale AW. Testing the Proposed Causal Link Between Cosmic Rays and Cloud Cover. Environmental Research Letters 2008;3 024001.

[89] Solanki SK, Krivova NA. Can Solar variability Explain Global Warming Since 1970? Journal of Geophysical Research 2003;108(A5) 1200-1212.

[90] Sugihara G, May RM. Nonlinear Forecasting as a Way of Distinguishing Chaos from Measurement Error in Time Series. Nature 1990;344 734-741.

[91] Tuovinen M, Mc Carroll D, Grudd H, Jalkanen R, Los S. Spatial and Temporal Stability of the Climatic Signal in Northern Fennoscandian Pine Tree-Ring Width and Maximum Density. Boreas 2009;38(1) 1-12.

[92] von Storch H, Zorita E, Jones PD, Dimitriev Y, González-Rouco F, Tett SF. Reconstructing Past Climate From Noisy Data. Science 2004;306(5696) 679-682.

[93] von Storch H, Zorita E. Comment on "Hockey Sticks, Principal Components, and Spurious Significance'" by S. McIntyre and R. McKitrick. Geophysical Research Letters 2005;32 L20701, doi:10.1029/2005GL022753,

[94] The impact of atmospheric carbon dioxide increasing on climate. Proceedings of the Soviet-American Workshop on Atmospheric Carbon Dioxide Increasing Study, Leningrad, 15-20 June 1981. Leningrad: Gidrometeoizdat; 1982 (in Russian).

[95] Wilson R, D'Arrigo R, Buckley B, Büntgen U, Esper J, Frank D, Luckman B, Payette S, Vose R, Youngblut D. A matter of divergence — tracking recent warming at hemispheric scales using tree-ring data. Journal of Geophysical Research 2007;112 D17103.1–D17103.17

Chapter 5

Some Indicators of Interannual Rainfall Variability in Patagonia (Argentina)

Marcela Hebe González

Additional information is available at the end of the chapter

1. Introduction

Argentina is located in the southeastern continental extreme of South America. The Andes Mountain Range extends 7240km all along the western region of the country with a mean height of 3660m above mean sea level (a.m.s.l.). Two topographical regions can be distinguished in the country: north of 40°S, the range is high and dense preventing the humidity access from the Pacific Ocean. As a result, atmospheric flow is dominated by the South Atlantic High and winds prevail from the northeast. An intermittent low pressure system, whose likely origin is a combination of thermal and dynamical effects, is located between 20º and 30ºS, in a dry and relatively high area east of the Andes. This system is observed throughout the year, but it is deeper in summer than in winter. When this low is present, northerly flow is favored at low levels over the subtropical region. Therefore, the water vapor entering at low levels comes either from the tropical continent or from the Atlantic Ocean. In the first case, the easterly low-level flow at low latitudes is channeled towards the south between the Bolivian Plateau and the Brazilian Planalto, advecting warm and humid air to southern Brazil, Paraguay, Uruguay and subtropical Argentina. South of 40ºS, the mean flow is from the west during all months and there are intermittent interruptions of polar fronts from the southwest associated with the displacement of Rossby waves over the Pacific ocean [1]. Storms occur frequently and erosion is one of the main features of the region.

The Argentinean Patagonia region is located between 30° and 55°S, where the wind circulation is predominantly westerly all year round, being more intense during winter. As a consequence of the interaction between wind patterns and orography, maximum precipitation (approximately 1600 mm/year in northern Patagonia) occurs in the vicinity of the mountains; decreasing eastward over Patagonian region especially in winter [2, 3] to a low of 300-500 mm/year. Consequently, eastern Patagonia is arid to semi-arid with hydrographic network that begins generally in the Andes Mountains and flows towards the Atlantic Ocean. Important rivers include the Negro, the Neuquen and the Limay river basins. The most favorable conditions for precipitation in the Patagonian Mountains and its eastern slopes are stationary fronts especially when the cold side of the anticyclone brings moist winds from the Atlantic. The result is stratiform clouds and extended rainfall. In few cases, rainfall occurs due to blocking high pressure systems located in Patagonia or the adjacent Atlantic resulting in extended periods with cloudiness and precipitation.

Neither the spatial nor temporal variability of rainfall in Patagonia have been widely studied, probably due to the limited amount of data that exists in this sparsely populated area with few measuring stations. Using linear regression adjustment to annual rainfall for the period 1950-1999, Castañeda and González [4] detected positive trends with 95% significance level on the order of about 2,5 mm/year in the northern and 1,23 mm/year in southern Patagonia, while precipitation decreased in the western and central zones. The authors used an alternative nonlinear methodology to detect a number of statistically significant breakpoints in order to identify homogeneous periods of stationary rainfall. This method is based on adjusting the data with consecutive linear segments, between periods with significant trends. Their results show that northeastern Patagonia experienced an abrupt change in the mean annual rainfall in the mid-to late 1960s. In the northwest, however, this change occurred later, in the mid-to late 1970s. Barros and Mattio [5] and Barros and Rodriguez Sero [6] analyzed long-term changes in the precipitation over the northern plateau of the Patagonia, especially during the 1940's. They found positive annual rainfall anomalies in northern and central Patagonia (Chubut province) especially in the 1940s and near the Andes Mountains in central Patagonia during the period 1920-1965. However, they detected negative annual rainfall anomalies in southern Patagonia (South of Santa Cruz province) in the 1950s. Minetti et al [7] performed an analysis of nonlinear trends in precipitation over Argentina and Chile using polynomial functions and spectral estimations. They showed that increasing quasi-linear trends were encountered all over Patagonia in the last century with no sign of stabilization in the average value while west of the Andes, in Chile, decreasing trends were observed. Other authors have studied sub-regions within Patagonia. For example, Russian et al [8] studied the relationship between interannual rainfall variability in Northern Patagonia, and tropospheric circulation. The authors detected statistical significant ($\alpha=0,05$) connections with El Niño/Southern Oscillation (ENSO) and Southern Annular Mode (SAM).

In contrast, the Comahue region, in northwestern Patagonia has been extensively studied because of the economic significance of the hydroelectric power stations operating in the region. Gonzalez and Vera [9] and Gonzalez et al [10] analyzed the connection between Comahue annual rainfall amounts and circulation patterns. The analysis for the Limay River Basin showed that the most important source of predictability came from the interannual variability of surface temperature in the tropical Indian Ocean and the ENSO phase while in the Neuquen River Basin geopotential heights at low levels in the Pacific Ocean, associated with the Rossby wave train that extend along the South Pacific, resulted in the best indicator of rainfall variability. Gonzalez and Cariaga [11] have developed a scheme prediction for winter and spring rainfall in the Comahue region in order to investigate the association between rainfall and various circulation patterns. The model showed that sea surface temperature (SST) in the tropical Indian Ocean, the wave train over the Pacific Ocean and the ENSO phase observed in the previous month collectively explain 66,5% of the variability of winter precipitation. A similar model was built for forecasting spring rainfall, showing that the dynamical systems are the main factors that contribute to generate spring precipitation accounting for 30% of spring rainfall variance.

Despite such studies, much remains that is unknown or poorly understood about the inter-annual rainfall variability in Patagonia. For example, the extent to which SSTs influence the annual rainfall remains unclear. The objective of this study therefore is to better understand the connection between the dynamics of the large scale atmospheric forcing, primarily atmospheric circulation patterns and sea surface temperature (SST) conditions and rainfall in Patagonian region.

Slow variations in the earth's boundary conditions (i.e. sea surface temperature) can influence global atmospheric circulation and thus, precipitation. A warming or cooling of some region of the oceans can act as a remote forcing generating teleconnections. Indeed, the most relevant SST pattern in the Pacific Ocean is the El Niño-Southern Oscillation (ENSO). The SST anomalies in tropical Pacific generate a Rossby wave pattern which propagates meridionally towards the middle-latitudes from the tropical source [12, 13, and 14]. This pattern, called the "Pacific South American Pattern", is described by Mo [1]. Some authors [15, 16], have studied the relation between above normal rainfall and "El Niño" events in northeastern Argentina. The authors found that Southern Brazil presents the strongest average signal in El Niño events. They showed that the general behavior toward opposite signals in the precipitation and circulation anomalies over Southern South America during almost the same periods of the El Niño and La Niña events indicates a large degree of linearity in the response to these events.

Related to SST anomalies but located in the Indian Ocean, is the "Indian Ocean Dipole" (IOD) [17]. A positive IOD period is characterized by cooler than normal water in the tropical eastern Indian Ocean and warmer than normal water in the tropical western Indian Ocean and it has been associated with decreased rainfall in central and southern Australia. For South America, Chan [18] showed that IOD excites a dipolar pattern in rainfall anomalies between subtropical La Plata basin and central Brazil where rainfall is reduced (enhanced) over the latter (former) during austral spring. It is also associated with a Rossby wave pattern extending from the subtropical south Indian Ocean to the subtropical South Atlantic. Liu [19] found evidence for the teleconnection detected between IOD and rainfall in Southern South America using the theory of planetary waves [20] and showed that the energy propagation path of planetary waves is approximately along the path of Rossby wave train, a possible dynamic explanation for such teleconnection pattern.

The relative intensity of the sub-tropical high and sub-polar low belts determines the intensity of westerlies over the Pacific Ocean and this factor highly influences rainfall regime in Patagonia. Another teleconnection that influences Patagonian rainfall is the Antarctic Oscillation, an annular-like pattern called "Southern Annular Mode" (SAM) [21]. The positive phase of the SAM is defined by negative pressure anomalies at high latitudes (higher than 65ºS) combined with wave-like pattern in the middle-latitudes. This feature increases zonal winds, decreases heat exchange between poles and mid-latitudes and so modifies storm tracks. Previous papers have shown SAM's influence on rainfall variability in some regions of South America. For example, Silvestri and Vera [22] found significant ($\alpha=0,05$) relation between SAM and rainfall amount in southeastern South America particularly during November and December. The authors found that the AAO influence was particularly strong during winter and late spring although of opposite sign and AAO positive (negative) phases were associated with the intensification of an upper-level anticyclonic (cyclonic) anomaly, weakened (enhanced) moisture convergence and decreased (increased) precipitation over southern South America. Reboita [23] detected a decrease of frontal activity when SAM is in a positive phase. These findings are consistent with the results of other investigators. For example, Zheng and Frederiksen [24] showed that the SAM affects summer rainfall variability in the New Zealand sector. They found New Zealand was drier especially in the South Island, when the 500-hPa height pattern had an anomalous high centered well south of New Zealand, with positive anomalies extending over much of the country. Reason and Rouault [25] showed that wetter (drier) winters in western South Africa occur during the negative (positive) SAM phase. The combined effect of the pressure patterns over the oceans and the SST anomalies in close proximity to the continent is another precipitation forcing. When SST is high, evaporation over the ocean is enhanced and moist onshore winds became intensified. Then, advection of moist air from the Atlantic Ocean intensifies and the air above the continent is more likely to generate precipitation. Moreover, significant SST anomalies over the oceans can modify the trajectories of pressure systems from the west.

The objective of this chapter is to better understand the possible global circulation patterns that influence seasonal precipitation in the Patagonian region of Argentina. A second objective is to determine the relationships between regional patterns, like the surrounding oceans SST or wind circulation and seasonal rainfall. This chapter is organized as follows: Section 2 describes the dataset and the methodology; Section 3 presents the results and section 4 discusses the mayor major findings and conclusions.

2. Data and methodology

Monthly rainfall data at 19 stations in Patagonia for the period 1981-2010 are used in this study. The data are derived from different sources including the National Meteorological Service (SMN) of Argentina, the Secretariat of Hydrology of Argentina (SRH) and Territorial Authority of the Limay, Neuquen and Negro rivers basins (AIC). The area of study is located between 37°S and 55°S and between 72ºW and 63ºW (Figure 1). All the selected stations have less than 20% of missing monthly rainfall data and their quality has been carefully proved. Some techniques were applied with that purpose: we discriminated cases with no precipitation in one month from missing data, no stations have records affected by changes of location and instrumentation. Rainfall greater that percentile 95 was controlled in order to detect outliers. In a consistency check, the monthly time series of observed rainfall at each station was compared with a nearby station, using double mass curve analysis, to see if it was physically or climatologically consistent. Suspicious observations according to inconsistencies were not considered. Seasonal rainfall is calculated as the accumulated precipitation in summer (December-January-February, DJF), autumn (March-April-May, MAM), winter (June-July-August, JJA) and spring (September-October-November, SON).

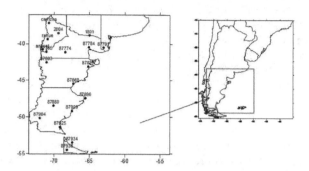

Figure 1. Stations used in the study

The following paragraphs describe the indexes used to analyze the relationship between seasonal rainfall and the different forcings described above.

El Niño-Southern Oscillation effect is evaluated using the mean SST in the EN3.4 region (ENSO) and they are obtained from the Climate Prediction Center (CPC, http://www.cpc.ncep.noaa.gov/products/) from National Oceanic and Atmospheric Administration (NOAA).

The IOD is commonly measured by an index defined as the difference between SST in the western (50°E to 70°E and 10°S to 10°N) and eastern (90°E to 110°E and 10°S to 0°S) equatorial Indian Ocean. The index is called the Dipole Mode Index (DMI) [17]. Data are obtained from SST DMI dataset derived from HadlSST dataset (http://www.jamstec.go.jp/frcgc/research/d1/iod/DATA/dmi_HadISST.txt)

The SAM pattern is represented quantitatively by an index called Antarctic Oscillation (AAO), defined as the difference in the normalized monthly zonal mean sea level pressure between 40°S and 65°S [21]. This index is obtained from http://ljp.lasg.ac.cn/dct/page/65572

To analyze the influence of the surrounding Atlantic and Pacific Ocean SST, five indices are defined (figure 2) as: mean SST in (38°S to 47°S and 68°W to 60°W) (S1), mean SST in (47°S to 55°S and 68°W to 60°W) (S2), both in Atlantic Ocean; mean SST in (35°S to 43°S and 71°W to 78°W) (S3), mean SST in (43°S to 55°S and 71°W to 78°W) (S4) in the Pacific Ocean coast. Finally, mean SST in (55°S to 70°S and 78°W to 60°W) (S5) in ocean between Argentina and Antarctic is defined.

Additionally, the 500 and 1000 Hpa geopotential height data are used to define the following indices (see also figure 3):

GH1: mean 500 Hpa geopotential height field in (30° to 40°S and 120°W to 80°W) in the South Pacific Ocean, representative of the intensity of sub-tropical high.

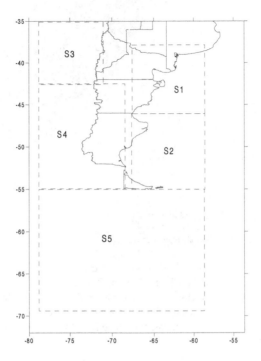

Figure 2. Definition of SST indexes.

GH2: mean 500 Hpa geopotential height field in (60º to 70ºS and 120ºW to 80ºW) in the South Pacific Ocean, representative of the intensity of sub-polar low belt

GH3: mean 500 Hpa geopotential height field in (36º to 55ºS and 65ºW to 55ºW) in the surrounding South Atlantic Ocean

GH4: mean 500 Hpa geopotential height field in (36º to 55ºS and 72ºW to 65ºW) over Patagonian region

GH5: mean 500 Hpa geopotential height field in (36º to 55ºS and 72ºW to 82ºW) in the surrounding South Pacific Ocean

The same indices are defined using 1000 Hpa geopotential heights and the same results are obtained, so they will not be detailed in this paper. SST and geopotential height data are from NCEP reanalysis.

Simultaneous correlations between seasonal (DJF, MAM, JJA and SON) rainfall in the 19 stations and indices defined below (ENSO, AAO, DMI, S1, S2, S3, S4, S5, GH1, GH2, GH3, GH4 and GH5) are calculated. Correlations greater than 0,37 (0,31) are significant at 95% (90%) confidence level. The results are used to describe the relationship between seasonal rainfall and circulation forcing.

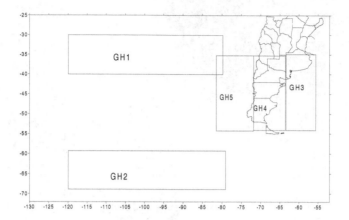

Figure 3. Definition of 500 Hpa Geopotential Height indexes.

The leading patterns of rainfall variability on seasonal time scales were obtained through a Principal Component Analysis (PCA), in the T-mode [26]. Therefore, the principal scores (PC) describe the spatial patterns while the loadings provide the temporal evolution of the patterns. The significant PCs have been detected using the Kaiser criterion [26] in which the eigenvalue must be greater than 1 in order to be considered. The resulting patterns were compared with correlation maps detailed above.

3. Results

The following figures show the correlation fields between seasonal rainfall and all the defined indices. Figures 4 and 5 show simultaneous correlations between JJA rainfall and defined indices. No significant correlations between rainfall and SST are observed. The only exception is a negative significant correlation between JJA rainfall and S5 (figure 4e) is present in a small area in eastern Patagonia, suggesting that cold waters south of the continent could displace systems trajectories towards the north over the continental region. Winter rainfall is increased by the positive phase of ENSO in the eastern of Patagonia (figure 4f). This result agrees with other authors who have investigated the impacts of "El Niño" in South America [14, 15, 16]. Figure 6 shows the temporal pattern of JJA rainfall at station 87860 (located in eastern Patagonia, see figure 1) and ENSO (figure 6a) and S5 (Figure 6b). Figures 5a and 5b show that JJA rainfall is enhanced especially in central and eastern Patagonia, when sub-tropical high and sub-polar lows are weakened (lower sub-tropical high and higher sub-polar low). So, the western flow weakens too and heat exchange between high and middle latitudes increases. This fact facilitates the meridional displacement towards the north of the precipitation systems associated with Rossby waves from the Pacific Ocean [1, 12, 13]. This result agrees with that of Aravena and Luckman [27]. Winter rainfall also increases in the western part of the study region with negative anomalies of geopotential heights over Patagonia (figure 5d) and surrounding Pacific Ocean (figure 5e), showing the influence of low systems displacing over the continent from the Pacific Ocean. Figure 6c shows the temporal behavior of JJA rainfall in Angostura in northwestern Patagonia (see figure 1) and GH5. Meanwhile, positive anomalies of geopotential heights over Atlantic Ocean coast (figure 5c) tend to increase rainfall in northeastern Patagonia, as indicated by statistically significant ($\alpha=0.05$) correlation centered on (39ºS; 64ºW), probably because of the increase of moist flow from the Atlantic when an anticyclonic system is present in that ocean.

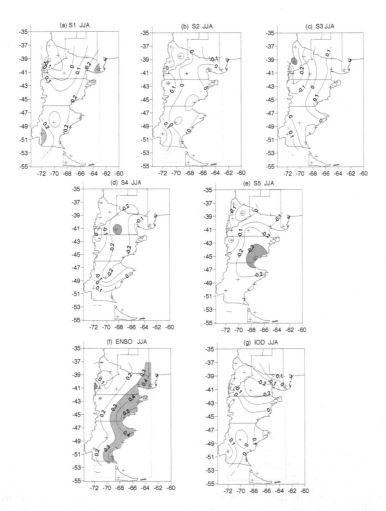

Figure 4. Correlation between JJA rainfall and S1 (a), S2 (b), S3 (c), S4 (d), S5 (e), ENSO (f) and IOD (g) as defined in the text. Areas with positive (negative) significant correlations are red (blue). Correlations greater than 0,31 (0,37) are significant at 90 (95)% confidence level. Stations were plotted.

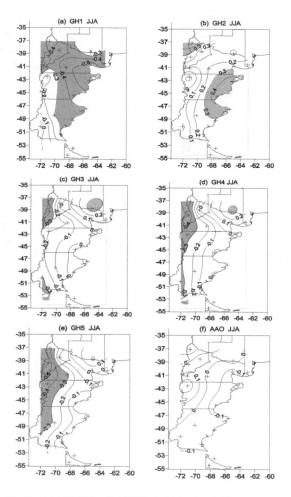

Figure 5. Idem Figure 4 for GH1 (a), GH2 (b), GH (c), GH4 (d), GH5 (e) and AAO (f).

Figures 7 and 8 show simultaneous correlations between SON rainfall and defined indices. The result show that western Patagonia spring rainfall tends to be increased by cold waters in the Atlantic Ocean coast (figure 7 a and 7b) and in the south of the continent (figure 7e). Cold waters over this part of the ocean probably causes systems to displace towards the northeast over the continent as a result of the continental warming which provides the energy necessary for its development. Another factor that influences spring rainfall in northwestern Patagonia is a positive phase of IOD (figure 7g), acting as a source of Rossby waves in the Indian ocean and a warm phase of ENSO (figure 7f). SON rainfall is enhanced by weakened sub-tropical highs (figure 8a) and sub-polar lows (figure 8b) in western and southern Patagonia. This fact implies that rainfall in that area is favored by negative phase of AAO (figure 8f). Besides, more

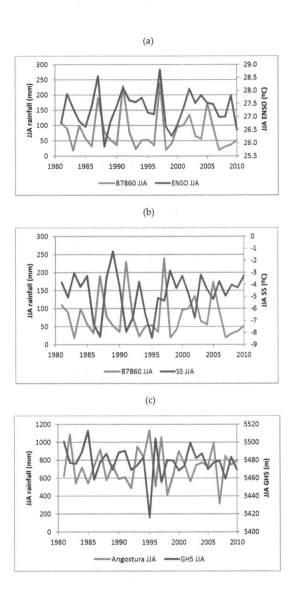

Figure 6. Temporal patterns of (a) JJA rainfall in 87860 station and ENSO, (b) JJA rainfall in 87860 station and S5 and (c) JJA rainfall in Angostura station and GH5.

rainfall is observed when cyclonic anomalies are present over Atlantic (figure 8c) and Pacific (figure 8e) coasts and over the continent (figure 8d). Figure 9 shows temporal patterns of SON rainfall in western Patagonia (87803 station, see figure 1) and ENSO (figure 9a), S5 (figure 9b) and GH5 (figure 9c).

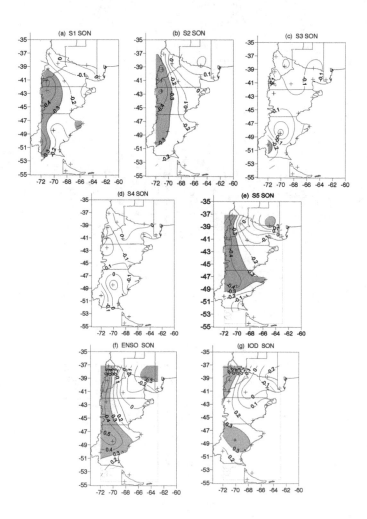

Figure 7. Idem figure 4 for SON.

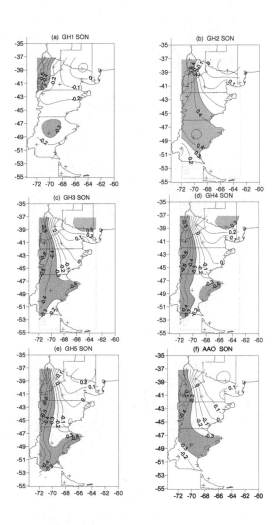

Figure 8. Idem figure 5 for SON.

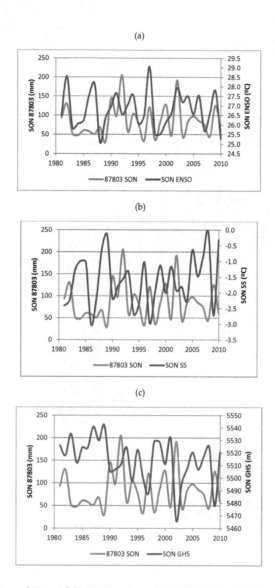

Figure 9. Temporal patterns of SON rainfall in 87803 station and ENSO (a), S5 (b) and GH5 (c).

Figures 10 and 11 show simultaneous correlations between MAM rainfall and defined indices. No significant correlation is detected between rainfall and sea surface temperature in the Atlantic Ocean (figures 10a and 10b). Warm waters in the Pacific coast (figures 10c and 10d) and a positive phase of ENSO (figure 10f) are related to greater than normal autumn rainfall

all over Patagonia. Figure 12a shows the time evolution of MAM rainfall in central Patagonia (87774 station, see figure 1) and S5. As in the case of winter and spring rainfall, weakened subtropical highs (figure 11a) and sub-polar lows (figure 11b) and therefore negative phase of AAO (figure 11f) are associated with enhanced rainfall in most of Patagonia. Besides, autumn rainfall in the northwest (Comahue region) is enhanced by cyclonic anomalies in the Atlantic (figure 11c), Pacific (figure 11e) and continental (figure 11d) sectors. Figure 12b shows the temporal pattern of MAM rainfall in northwestern Patagonia (Angostura station, see figure 1) and GH5.

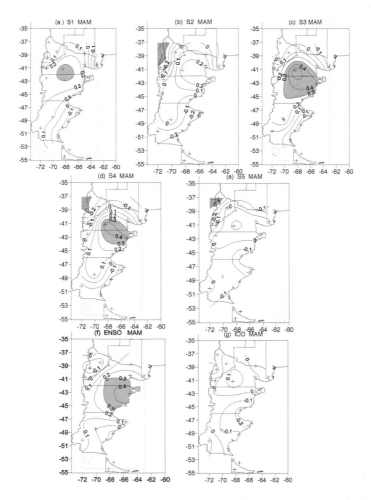

Figure 10. Idem figure 4 for MAM.

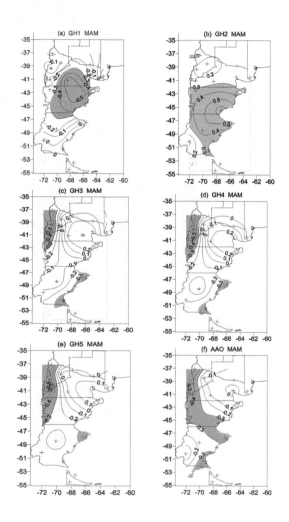

Figure 11. Idem figure 5 for MAM.

(a)

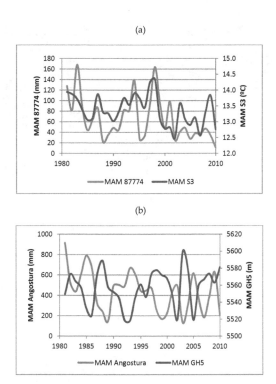

(b)

Figure 12. Temporal patterns of (a) MAM rainfall in 87774 station and S3 and (b) MAM rainfall in Angostura station and GH5.

Figures 13 and 14 show simultaneous correlations between DJF rainfall and defined indices. Summer rainfall is very scarce all over Patagonia. No influence is observed between summer precipitation and SST (figure 13). Only cyclonic anomalies over the Atlantic (figure 14c) and Pacific (figure 14e) surrounding oceans and over the continent (figure 14d) are related to enhanced rainfall in the west of Patagonia. Figure 15 shows the time evolution of DJF rainfall in western Patagonia (87803 station, see figure 1) and GH5.

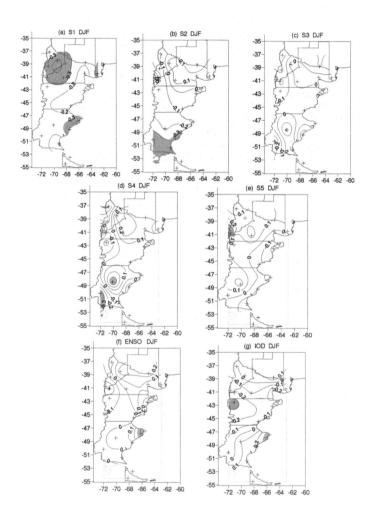

Figure 13. Idem figure 4 for DJF.

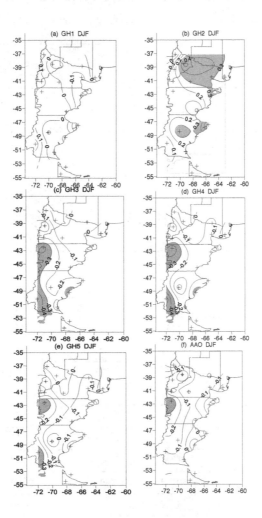

Figure 14. Idem figure 5 for DJF.

In an attempt to reproduce the patterns identified by the indices tested, a PCA analysis was applied to accumulated seasonal rainfall anomalies. The first four spatial patterns of rainfall anomaly are detailed in figures 16 (JJA), 17 (SON), 18 (MAM) and 19 (DJF). Table 1 details the variance explained by each significant pattern. Tables 2, 3, 4 and 5 detailed the correlations between each eigenvector and indices defined in the text for JJA, SON, MAM and DJF, respectively.

Number of loading	JJA	SON	MAM	DJF
1	43,9	39,3	43,6	22,7
2	19,9	15,6	13,2	18,4
3	13,1	10,5	11,1	14,1
4	6,3	7,3	6,3	9,2
5	5	5,8	6,1	8,1
6		4,9	4,2	5,3
7		3,7	3,7	4,8
8				4,1
9				3,5
Total variance explained by significant patterns.	88,2	87,1	88,2	90,1

Table 1. Variance explained by significant loadings (%)

INDICES	Ei 1	Ei 2	Ei 3	Ei 4
EN34	0,135	0,559	0,158	0,086
AAO	0,173	-0,174	-0,014	0,275
IOD	0,420	0,358	0,035	0,247
S1	-0,009	0,197	0,407	0,144
S2	0,109	-0,171	0,317	0,220
S3	-0,274	0,149	0,179	-0,195
S4	0,046	-0,242	0,158	0,034
S5	-0,005	-0,435	-0,119	0,036
GH1	-0,323	-0,435	0,091	-0,163
GH2	0,079	0,499	-0,033	-0,153
GH3	-0,308	0,046	-0,321	-0,201
GH4	-0,403	0,033	-0,351	-0,332
GH5	-0,520	0,009	-0,315	-0,428

Table 2. Correlation between indices defined in the text with the first four temporal patterns (eigenvectors 1, 2, 3 and 4) derived from PCA in JJA. Correlations significant at α=0.05 are shaded.

INDICES	Ei 1	Ei 2	Ei 3	Ei 4
EN34	-0,615	0,238	0,250	-0,271
AAO	0,509	0,055	-0,086	0,046
IOD	-0,432	0,421	0,144	-0,169
S1	0,287	0,135	-0,182	-0,298
S2	0,287	-0,074	-0,337	-0,172
S3	-0,002	-0,103	0,012	-0,364
S4	-0,062	0,000	-0,343	-0,369
S5	0,558	-0,160	-0,332	0,218
GH1	0,559	-0,276	0,064	0,197
GH2	-0,628	0,181	0,084	-0,120
GH3	0,646	0,028	-0,317	-0,121
GH4	0,644	0,011	-0,265	-0,232
GH5	0,665	-0,104	-0,125	-0,282

Table 3. Idem Table 2 for SON

INDICES	Ei 1	Ei 2	Ei 3	Ei 4
EN34	0,038	-0,074	-0,205	-0,383
AAO	0,438	0,110	-0,071	0,334
IOD	0,026	0,020	-0,095	-0,092
S1	-0,217	0,026	-0,272	-0,253
S2	0,244	0,138	-0,168	0,056
S3	-0,123	0,217	-0,261	-0,238
S4	0,212	0,212	-0,103	-0,104
S5	0,181	0,278	-0,044	0,352
GH1	0,208	0,087	0,249	0,403
GH2	-0,144	-0,189	-0,157	-0,399
GH3	0,511	-0,042	-0,149	0,096
GH4	0,596	0,084	-0,105	0,095
GH5	0,679	0,106	-0,009	0,117

Table 4. Idem Table 2 for MAM

INDICES	Ei 1	Ei 2	Ei 3	Ei 4
EN34	0,097	0,280	0,055	0,087
AAO	0,105	0,198	0,224	-0,179
IOD	0,033	0,138	0,330	0,142
S1	0,044	0,002	0,036	-0,237
S2	-0,292	0,098	-0,218	0,144
S3	0,007	0,022	-0,055	0,075
S4	-0,266	0,070	-0,082	0,281
S5	-0,498	-0,068	0,272	0,288
GH1	-0,015	-0,093	-0,018	0,016
GH2	-0,209	0,264	0,021	0,253
GH3	-0,030	-0,013	-0,264	-0,245
GH4	-0,017	-0,113	-0,184	-0,174
GH5	0,017	-0,146	-0,127	-0,114

Table 5. Idem Table 2 for DJF

GH1	-0,015	-0,093	-0,018	0,016
GH2	-0,209	0,264	0,021	0,253
GH3	-0,030	-0,013	-0,264	-0,245
GH4	-0,017	-0,113	-0,184	-0,174
GH5	0,017	-0,146	-0,127	-0,114

Table 5.   Idem Table 2 for DJF

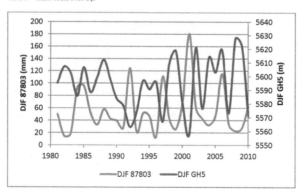

Figure 15. Temporal patterns of DJM rainfall in 87803 station and GH5.

Figure 15. Temporal patterns of DJM rainfall in 87803 station and GH5.

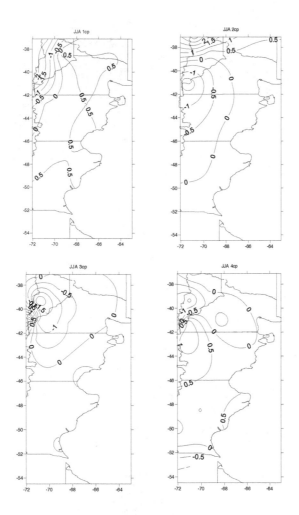

Figure 16. First four spatial patterns of JJA rainfall anomaly derived from CPA analysis during 1981-2010 period.

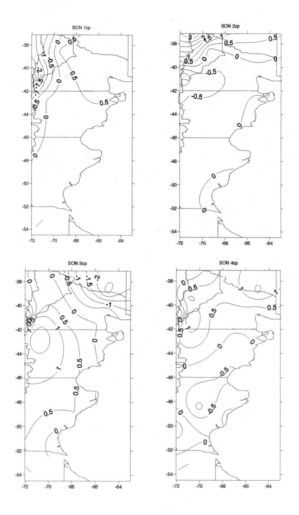

Figure 17. Idem Figure 16 for SON

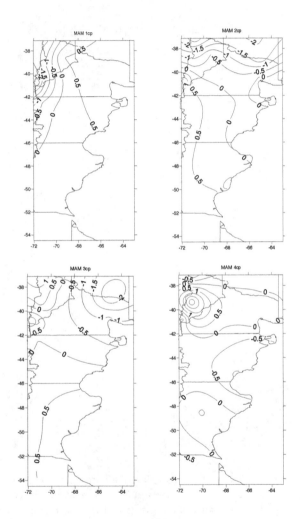

Figure 18. Idem Figure 16 for MAM

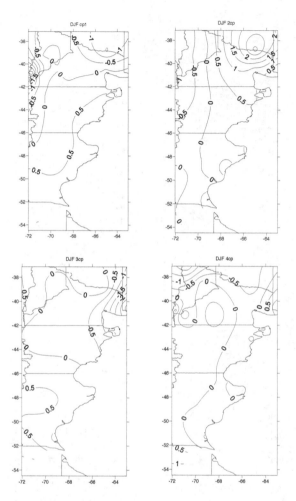

Figure 19. Idem Figure 16 for DJF

Some similarities with the results previously described are detected when the correlations between the eigenvectors and indices are analyzed. In relation to the observed correlations in JJA (Table 2) one can see that the correlation between S5 and the second eigenvector (-0,435) reflects the fact that precipitation increases in eastern Patagonia (Figure 16, JJA 2cp) when S5 is low. Similarly, precipitation increases in this region during the warm phase of ENSO (correlation between ENSO and the second eigenvector 0,559), and when sub-tropical highs (correlation between GH1 and the second eigenvector -0,435) and sub-polar lows (correlation between GH2 and the second eigenvector 0,499) are weaken.

Table 3 shows that in SON, the correlation of the third eigenvector with S2 (-0,337) and S5 (-0,332) reflects the fact that in western Patagonia (Figure 17, 3cp) precipitation increases when S5 and S2 are low. The correlation between IOD and the second eigenvector (0,421) and the first eigenvector (-0,432) indicates that in western Patagonia (Figure 17, 1cp) and in north-western Patagonia (Figure 17, 2cp) precipitation decreases during negative phase IOD. Also the correlation between AAO and the first eigenvector (0,509) also indicates enhanced precipitation in western Patagonia during negative phase of AAO. Sub-tropical highs (correlation between GH1 and the first eigenvector 0,559) and the sub-polar lows (correlation between GH2 and the first eigenvector -0,628) weaken and during the positive phase of ENSO (correlation between first eigenvector and ENSO -0,615), indicate an increase in precipitation in western Patagonia (figure 17, 1CP).

Table 4 shows that the correlation between fourth eigenvector with AAO in MAM is 0.334 and this indicates that in central-eastern Patagonia (Figure 18, 4CP) enhanced precipitation occurs during the negative phase of AAO. Furthermore, when sub-tropical highs (correlation between GH1 and the fourth eigenvector = 0.403) and the sub-polar lows (correlation between GH2 and the fourth eigenvector = -0.399) weakens, enhanced precipitation occurs in central Patagonia (Figure 18, 4CP).

No significant correlation was found in DJF probably because rainfall is scarce in the Patagonia Plateau and the greatest precipitation occurs in winter, near the Andes Mountains.

The Patagonia region in Argentina is of particular interest and is the source of water for hydropower generation for the national grid, the local subsistence economies, agriculture (fruit production) and mining. This complexity is expressed in social and environmental exploitation processes, inequitable distribution of access-control of natural resources, in population displacement and urban growth. Recent climate studies indicate potential future increase in water stress in Patagonia region in Argentina, which will affect the ecological productivity and ecosystem services [28]. Consequently, it is of particular interest to identify current trends and anticipate future scenarios that might lead to environmental and social conflicts. From a climatic point of view, many authors have studied this issue in the South American region [29,30] and found significant changes in precipitation since the beginning of the century. But there are indications of changes in rainfall trends during the past 30 years in some regions [31]. As low frequency variabilities are fed by individual events that occur every year, there is an urgent need for the study of the interannual variability as well as the potential for predicting seasonal rainfall based on teleconnection indices.

4. Conclusions

This findings reported in this chapter are a preliminary attempts to identify the main circulation patterns related to precipitation variability in the Patagonian region of Argentina. The chapter analyzed the relative significance of the various teleconnections responsible for precipitation variability. The teleconnections analyzed include ENSO, AAO, IOD, geopotential heights and SST patterns. The principal method of analysis was the correlation and the PCA applied to seasonal rainfall.

The results show that the factors which affect precipitation most highly depend on the season and the region. Cyclonic anomalies over the surrounding Atlantic and Pacific Oceans and over the continent enhance precipitation all over Patagonia and in all seasons, although in winter the Atlantic influence is less. The weakening of sub-tropical hights and sub-polar lows over the Pacific Ocean is associated with above normal precipitation throughout Patagonia in transition seasons (MAM and SON) in the eastern part in winter.

Some influences of the SST patterns are present. Warm phase of ENSO enhances rainfall in eastern Patagonia in autumn and winter and in western and southern Patagonia in spring. Warm water in the Pacific coasts tends to increase rainfall in central and eastern Patagonia in autumn while cold Atlantic coast enhances precipitation in west and south Patagonia in spring. The IOD signal is present only in autumn in the western and southern parts of the study region.

Future analysis will investigate the relative contribution of each forcing to seasonal rainfall but that objective is beyond the scope of this paper.

Acknowledgements

Rainfall data were provided by the Territorial Authority of the Limay, Neuquen and Negro rivers basins (AIC), the National Meteorological Service of Argentina (SMN) and the Secretariat of Hydrology of Argentina (SRH). This research was supported by UBACyT 2010-2012 CC02, UBACyT 2011-2014 01/Y128 and CONICET PIP 112-200801-00195, PICT 2010-2110.

Author details

Marcela Hebe González

Address all correspondence to: gonzalez@cima.fcen.uba.ar

Department of Atmospheric and Oceanic Science - FCEN-University of Buenos Aires, Research Center of Ocean and Atmosphere – CONICET/UBA; UMI-IFAECI/CNRS, CIMA - 2º piso, Pabellón II, Ciudad Universitaria, Ciudad Autónoma de Buenos Aires, Argentina

References

[1] Mo, K. C. Relationships between low frequency variability in the Southern Hemisphere and sea surface temperature anomalies. J. Climate (2000). , 13-3599.

[2] Prohaska, F. J. Climates of Central and South America". In: World Survey of Climatology. Amsterdam: Elsevier Cientific Publishing Company; (1976). , 57-69.

[3] Russian, G, Agosta, E, & Compagnucci, R. H. Variabilidad interanual e interdecádica de la precipitación en la Patagonia norte. Geoacta (2010). , 35-27.

[4] Castañeda, M. E, & González, M. H. Some aspects related to precipitation variability in the Patagonia region in Southern South America. Atmósfera (2008). , 21(3), 303-317.

[5] Barros, V. R, & Mattio, H. F. Tendencias y fluctuaciones en las precipitaciones de la región patagónica. Meteorológica (1978). VIII-IX , 237-248.

[6] Barros, V R. and Rodriguez Sero JA. Estudio de las fluctuaciones y tendencias de la precipitación en el Chubut utilizando funciones ortogonales empíricas. GEOACTA (1979). , 10(1), 1979-204.

[7] Minetti, J. L, Vargas, W. M, Poblete, A. G, Acuña, L. R, & Casagrande, G. Non-linear trends and low frequency oscillations in annual precipitation over Argentina and Chile, Atmósfera (2003). , 1931-1999.

[8] Russian, G. F, Agosta, E. A, & Compagnucci, R. H. Circulación troposférica de gran escala asociada a la precipitacion en Patagonia norte. In Proceedings of CONGREMET XI, 28May-1June (2012). Mendoza, Argentina.

[9] González, M. H, & Vera, C. S. On the interannual winter rainfall variability in Southern Andes. International Journal of Climatology (2010). , 30-643.

[10] González, M. H, Skansi, M. M, & Losano, F. A statistical study of seasonal winter rainfall prediction in the Comahue region (Argentine). ATMOSFERA (2010). , 23(3), 277-294.

[11] González, M. H, & Cariaga, M. L. Estimating winter and spring rainfall in the Comahue region (Argentine) using statistical techniques. Advances in Environmental Research (2011). , 11, 103-118.

[12] Kidson, J. Principal modes of southern hemisphere low frequency variability obtained from NCEP-NCAR reanalyses. J. Climate (1999). , 1-1177.

[13] Nogues PaegleJ and Mo, KC. Linkages between Summer Rainfall Variability over South America and Sea Surface Temperature Anomalies. J. Climate (2002). , 15, 1389-1407.

[14] Ropelewski, C, & Halpert, M. Global and Regional scale precipitation patterns associated with El Niño. Mon Wea Rev (1987). , 110-1606.

[15] Grimm, A, Barros, V, & Doyle, M. Climate variability in Southern South America associated with El Niño and La Niña events. J. Climate (2002). , 13-35.

[16] Vera, C, Silvestri, G, Barros, V, & Carril, A. Differences in El Niño response in Southern Hemisphere. J. Climate (2004). , 17-9.

[17] Saji, N. H, Goswami, B. N, Vinayachandran, P. N, & Yamagata, T. A dipole mode in the tropical Indian Ocean. Nature 401; , 360-363.

[18] Chan, S, Behera, S. K, & Yamagata, T. Indian Ocean Dipole influence on South American rainfall. Geophysical Research Letter (2008). L14S12 DOPI: 10.1029/2008GL034204.

[19] Liu, N, Chen, H, & Lu, L. Teleconnection of IOD Signal in the Upper Troposphere over Southern High Latitudes. Journal of Oceanography (2007). , 63, 155-157.

[20] Hoskins, B. J, & Karoly, D. J. The steady linear response of a spherical atmosphere to thermal and orographic forcing. J. Atmos. Sci.(1981). , 38-1179.

[21] Thompson, D. W, & Wallace, J. M. Annular modes in the extratropical circulation. Part I: Month-to-month variability. J. Climate (2000). , 13, 1000-1016.

[22] Silvestri, G, & Vera, C. Antarctic Oscillation signal on precipitation anomalies over southeastern South America. Geophys Res Lett. (2003). , 30-21.

[23] Reboita, M. S, & Ambrizzi, T. and Da Rocha, R. Relationship between the Southern Annular Mode and Southers Hemisphere atmospheric systems. Revista Brasilera de Meteorologia (2009). , 24-1.

[24] Zheng, X, & Frederiksen, C. A study of predictable patterns for seasonal forecasting of New Zealand rainfall. J. Climate (2006). , 19-3320.

[25] Reason, C, & Rouault, M. Links between the Antartic Oscillation and winter rainfall over western South Africa. Geophys Res Lett. (2005). DOI:GL022419.

[26] Green, P. Analysing Multivariate Data. New York: Dryden Press; (1978).

[27] Aravena, J. C, & Luckman, B. H. Spatio-temporal patterns in Southern South America. Int. J. Climatol. (2008). DOI:joc.1761.

[28] Paruelo, J. Valoración de servicios ecosistémicos y planificación del uso del territorio: ¿es necesario hablar de dinero?. In: Laterra P, Jobbagy EG y Paruelo JM (ed.) Expansión e intensificación agrícola en Argentina: Valoración de bienes y servicios ecosistémicos para el ordenamiento territorial. Buenos Aires: INTA; (2012).

[29] Barros, V, Doyle, M, & Camilloni, I. Precipitation trends in southeastern South America: relationship with ENSO phases and the low-level circulation. Theoretical and Appl. Climatology (2008).

[30] Liebmann, B, Vera, C. S, Carvalho, L, Camilloni, I, Hoerling, M, Allured, D, Barros, V, Báez, J, & Bidegain, M. An Observed Trend in Central South American Precipitation. J. Climate (2004). , 17(22), 4357-4367.

[31] González, M. H, Dominguez, D, & Nuñez, M. N. Long term and interannual rainfall variability in Argentinean Chaco plain region. In: Martin O and Roberts T (ed.) Rainfall: Behavior, Forecasting and Distribution. New York: Nova Science Publishers Inc; (2012). , 1-21.

Chapter 6

A Review of Climate Signals as Predictors of Long-Term Hydro-Climatic Variability

Shahab Araghinejad and Ehsan Meidani

Additional information is available at the end of the chapter

1. Introduction

1.1. Large scale climate signals

In many parts of the world coupled oceans atmospheric phenomenon provide important predictive information about hydrologic variability. Therefore, studying the relationships of these large scale features of the atmosphere with hydroclimatic events is helpful in hydrological and meteorological long-lead forecasting, promoting awareness about climate variability, and water resources management. As the time scale over which the oceans respond are slower than atmosphere, efforts have been focused mostly on investigating the links of sea surface temperatures (SSTs) and sea level pressures (SLPs) of oceans with atmospheric changes for climate monitoring or use as potential hydro-climatic predictors.

This chapter reviews the characteristics of several widely known teleconnection indices and their effects on different regions of the world. The goal is to present basic information that might be useful for analysis and study of teleconnections. Knowledge concerning the contemporary dynamics of these teleconnections is essential contextual information against which the manifestations and impacts of future climate change can be assessed. For example, climate change is likely to change the intensity, timing, positional loci, as well as associated impacts of some of these teleconnections. The ability to assess such changes with a degree of accuracy is possible only if we have detailed information regarding past patterns of behavior.

2. El Niño Southern Oscillation (ENSO)

El Niño is one of the largest oscillations of the climate system and is defined as warmer than normal condition of Pacific Ocean surface temperature in tropical eastern parts (Figure

1).Under normal conditions, Pacific Ocean currents are east-to-west on the equator, causing the upwelling of deeper water toward the surface in the eastern parts. Every 2 to 7 years this westward flow weakens and reverses to an eastward flow of water called equatorial Kelvin waves. This abnormal condition is El Niño (warm phases) and typically can last for a duration of few months or more. La Niña episodes (cool phases) follows El Niño ones and alter the situation to normal and affect a global atmospheric pressure variation called the Southern Oscillation (SO), leading to the widely used term ENSO. Properties of ENSO are shown briefly in Table 1.

Climate signal	Region of occurrence	Indicator Index	Threshold	Phase persistence
ENSO	Equatorial Pacific Ocean	SOI	+/- 8	6-18 months
		Nino 1+2	+/- 0.5	
		Nino 3		
		Nino 3.4		
		Nino 4		

Table 1. The ENSO properties in brief

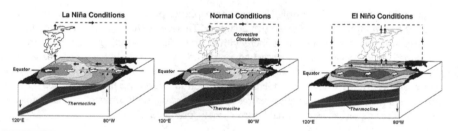

(Source:NOAA / PMEL / TAO Project Office, Dr. Michael J. McPhaden, Director, http://www.pmel.noaa.gov/tao/elni-no/el-nino-story.html)

Figure 1. Schematic diagram of ENSO climate pattern in the Pacific Ocean

ENSO is one of the most widely studied large-scale climatic variability that affects temperature and precipitation in regions all over the globe. Dettinger and Diaz (2000) showed that these large-scale tropical fluctuations yield similarly global-scale fluctuations on streamflow and that streamflow teleconnections are as pervasive as meteorological teleconnections. Based on studies of 1345 sites around the world, El Niño variations have been found to correlate with streamflow in many parts of the Americas, Europe and Australia (Dettinger and Diaz, 2000). The effects differ significantly by location, as abundant rains become scarce in some areas while other areas experience flooding (Figure 2). For instance, El Niño episodes are accompanied by reduction in winter and spring rainfall over much of eastern Australia and the higher latitudes and La Niña increases the probability of eastern and northern Australia being wetter than normal (Power et al. 2005).

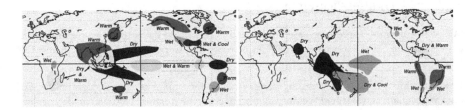

Figure 2. The regions having greatest impacts of El Niño during (*left*) December-February and (*right*) June-August. (Source: NOAA / PMEL / TAO Project Office, Dr. Michael J. McPhaden, Director, http://www.pmel.noaa.gov/tao/elnino/el-nino-story.html)

Currently, there is no single data set universally accepted for measurements of ENSO (Beebee and Manga 2004). Commonly used ENSO indices include regional SST indices (e.g., Nino-1+2, Nino-3, Nino-4, Nino-3.4) and SOI.

2.1. Nino index

These indices are based on the average sea surface temperature anomalies of the following regions and include, Niño 1+2 (0-10° S and 80-90° W), Niño 3 (5°N-5°S and 90-150°W), Niño 4 (5°N-5°S and 160°E-150°W) and Niño3.4 (5°N-5°S and 170-120°W) (Figure 3).

The raw values of the index are available from the Climate Prediction Center of The National Oceanic and Atmospheric Administration (NOAA) (http://www.cpc.ncep.noaa.gov/data/indices/) which is produced by using Reynolds OISST.v2 data from 1982 to present. When the index is positive then the temperature of the Pacific Ocean water is warmer than normal in the Nino regions and when the index is negative then water temperatures are cooler than normal. An El Niño or La Niña event is identified if, for example, the 3-month running-average of the NINO3.4 index exceeds +0.5°C for El Niño or -0.5°C for La Niña for at least 5 consecutive seasons. For example, El Niños occurred in 1982-83, 1986-87 and 1997-98 (Figure 4).

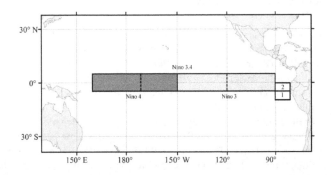

Figure 3. Pacific Ocean and corresponding Nino regions

2.2. Southern Oscillation Index (SOI)

SOI is an atmospheric pressure based index and is calculated using the pressure differences between Tahiti (17° 52' 0" south and 149° 56' 0" west) and Darwin (12° 27' 56" south and 130° 50' 33" east).

Positive values of SOI indicate strong trade winds in the tropics and the tropical Pacific. SOI data are available since late 19th century to present from the Australian Bureau of Meteorology website (http://www.bom.gov.au/climate/enso/).Sustained negative values of the SOI greater than −8 often indicate El Niño episodes and conversely, sustained positive values greater than +8 are typical of a La Niña episode. The 5-month running average of the SOI values also has been used as indicator of the episodes.

There are a few different methods for calculating the SOI. The method used by the Australian Bureau of Meteorology is the Troup SOI which is the standardized anomaly of the Mean Sea Level Pressure difference between Tahiti and Darwin. It is calculated as follows:

$$\text{SOI} = \frac{(P_{diff} - P_{diffav})}{SD(P_{diff})} \tag{1}$$

Where: P_{diff} is the average Tahiti MSLP (mean sea level pressure) for the month minus the average Darwin MSLP for the month; P_{diffav} is the long term average of P_{diff} for the month; and $SD\ (P_{diff})$ is the long term standard deviation of P_{diff} for the month.

While the formula is fairly standard, values are commonly multiplied by 10 to rescale data, resulting in a range from approximately -35 to 35. The SOI is usually computed on a monthly basis for a time period of a year or longer. The values of the index are obtained from the aforementioned sources and the fluctuation of the monthly values is demonstrated in Figure 4. As it is shown in the figure, Nino indices are very similar, while SOI tracks close to these indicators in an inverse form.

3. Pacific Decadal Oscillation (PDO)

PDO index was noted by fisheries scientist Steven Hare in 1996, based on observations of Pacific fisheries cycles and has often been described as a long-lived El Niño-like pattern of Pacific climate variability (Zhang et al. 1997). PDO is based on the monthly sea surface temperature variability of the Pacific Ocean north of 20°N on decadal scale, while the global average anomaly is subtracted from the SSTs to account for global warming (Mantua et al. 1997). Normally, only November to March values are used in calculating the PDO index because year to year fluctuations are most apparent during the winter months (Mantua and Hare, 2002). Even though the PDO and ENSO are related to the same ocean they have two important differences: on one hand 20th century PDO events have a persistence on the order of 20 to 30 years, while, typically ENSO persists for 6 to 18 months. Furthermore, the climate finger prints of the PDO are most visible in the north Pacific while the latter one exists in the tropic. Several studies have indicated two full phases of PDO in the past century (e.g., Mantua

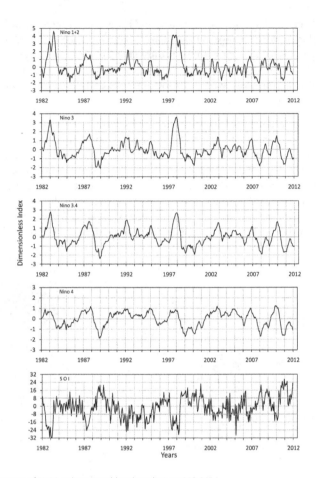

Figure 4. Fluctuation of ENSO indices monthly values during 1982-2011

et al. 1997 and Tootle et al. 2005): cool PDO regimes are reported from 1890-1924 and again from 1947-1976, while warm PDO regimes dominated from 1925-1946 and from 1977 through the mid-1990's. The threshold of cool and warm phases of PDO is zero.

During a positive (warm) phase of PDO the west Pacific is cooler than parts of the eastern Pacific. Opposite pattern occur in the negative or cool phases. These SST fluctuations affect climate patterns of vast regions but the results are more pronounced in the North Pacific. According to precipitation anomalies, positive (warm) phase of the PDO are negatively correlated, mostly, with eastern Australia, Korea, Japan, the Russian Far East and much of Central America (Figure 5) and positively correlated with southwest US. In terms of the temperature anomalies, positive (warm) phase of the PDO tends to coincide generally with anomalously warm temperatures in northwestern North America ($r^2>0.16$) and cool temperatures in eastern China and southeast US (Mantua and Hare, 2002). Pacific salmon production can also be a good indicator of change in PDO phase. Mantua et al. (1997) noted that for much of the past two decades (coincident with the positive PDO phase), salmon fishers in Alaska have prospered while those in the Pacific Northwest have suffered. Yet, in the 1960s and early 1970s, when the Pacific Northwest was under the influence of the negative PDO phase, their fortunes were essentially reversed.

Even in the absence of a comprehensive theoretical or mechanistic understanding, PDO index provides significant predictive information for improving long-term climate forecasts of the regions. This is true because of the PDO's strong tendency for multi-season and multi-year persistence. Gedalofand Smith (2001) applied tree-ring method and Biondi et al. (2001) used ring-widths from moisture stressed trees to reconstruct a PDO index extending back to 1600s. A good source for Monthly values of the index from 1900 to present is the Joint Institute Study of the Atmosphere and Ocean, University of Washington (http://jisao.washington.edu/pdo/). Figure 6 shows how the November-to-March average values for the PDO index have shifted phases and varied over the past decades.

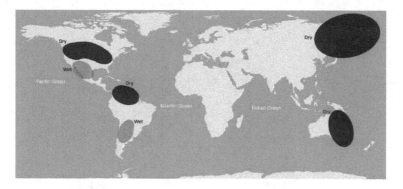

Figure 5. The regions having impacts of PDO

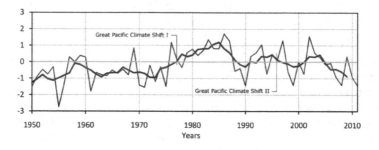

Figure 6. Fluctuation of the Nov-Mar averaged PDO index, with a 5-year running average (bold line).

4. North Atlantic Oscillation index (NAO)

The winter time station-based NAO index is the climate variability mode in North Atlantic Ocean and is defined as the difference in normalized mean winter (December to March) sea level pressure (SLP) anomalies between the *Stykkisholmur/Reykjavik*, Iceland and *Lisbon*, Portugal (Hurrell 1995). But this is not the only acceptable method for defining the index. Most modern NAO indices are based either on the simple difference in surface pressure anomalies between various northern and southern locations as Gibraltar and Reykjavik sites (Jones et al., 1997 and Brandimarte et al., 2011), or on a Principal Component (PC) time series approach (e.g., Portis et al.,2001 and Hurrell and Deser, 2009). The station-based method incorporates the simplicity of data construction into data availability for a period extending back to middle 19th century. A limitation of this approach is that it is unable to track the movement of the NAO centers of action through the annual cycle, owing to the spatial distribution and fixed nature of the station network used. Also, individual station pressure readings can be noisy due to small-scale and transient meteorological phenomena unrelated to the NAO. The PC-based method has the advantage of using gridded sea level pressure data and can be a better representation of the phenomenon according to its dimensions. On the other hand, this approach is limited to the recent decades as the remote sensed SST data became available for every grid of the oceans. More detailed discussions of issues related to the NAO indices are available in Hurrelland Deser (2009) and Hurrell et al. (2003).The threshold of warm and cool phase of NAO signal is zero and each phase might last several years.

NAO affects the strength and tracks of westerly winds and storms across the Atlantic and into Europe (particularly in winter) which in turn results in changes in temperature and precipitation patterns often extending from eastern North America to western and central Europe (Dettinger and Diaz, 2000). Westerly winds blowing across the Atlantic bring moist air into Europe. In NAO positive phases westerly winds are strong and lead to increased rainfall, cool summers and mild and wet winters across northern Europe, below average temperatures and precipitation in Greenland and oftentimes across southern Europe and the Middle East and to a lesser extent warmer winter over eastern North America and eastern South Canada. In contrast, if the index is negative, westerly winds are suppressed, the temperature in central and northern Europe and also southern United States is more extreme in summer and these areas suffer cold winters. Storms track southerly toward the Mediterranean Sea and this brings increased storm activity and rainfall to southern Europe and North Africa. Especially during the months of November to April, the NAO is responsible for much of the variability of weather in the North Atlantic region, affecting wind speed and wind direction changes, changes in temperature and moisture distribution and the intensity, number and track of storms.

The index shows annual variability but has the tendency to remain in single phase for intervals lasting several years (Hurrell 1995). Hurrell and van loon (1995) reported the NAO cool phase from 1952-1972 and 1977-1980 and also warm phases from1950-1951, 1973-1976, and 1981 to present. Winter time (December through March-JDFM), monthly, seasonal and annual NAO values are obtainable from the US National Center for Atmospheric Research (NCAR) (http:// www.cgd.ucar.edu/cas/jhurrell/indices.html).The anomalies of the Hurrel station-based and PC-based DJFM NAO indices are demonstrated in Figure 7.As shown in the figure, their behavior is very close to each other. In station-based method, the SLP anomalies at each station were normalized by dividing each seasonal mean pressure by the long-term mean (1864-1983) standard deviation.

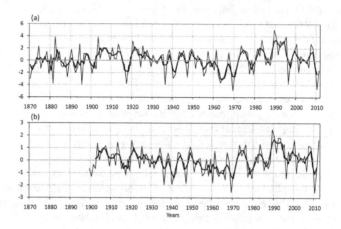

Figure 7. Fluctuation of (a) Hurrell station-based DJFM NAO index and (b) Hurrell PC time series of DJFM SLP over Atlantic sector (20-80°N, 90°W-40°E), with three-year running average (bold line).

5. Atlantic Multi-decadal Oscillation index (AMO)

The continuing sequences of long-duration changes in the de-trended sea surface temperature over the North Atlantic from 0-70°N are termed Atlantic Multi decadal Oscillation (AMO), with cool and warm phases that can last from 20-40 years at a time (Enfield et al. 2001).The time series of the index are calculated from the Kaplan SST dataset which is updated monthly and are created in two versions: smoothed and unsmoothed. The former version is smoothed with a 121 month smoother using earlier 61 and later 60 months values. While the observed AMO cycles are only available at most for the last 150 years, paleo-reconstructed climate data using methods such as tree rings and ice cores have shown that oscillations similar to those observed by instrument have been occurring for approximately the past five centuries (Gray et al., 2004). These large swings in North Atlantic SSTs are probably caused by natural internal variations in the strength of ocean thermohaline circulation and the associated meridional heat transport (Collins and Sinha, 2003). In the 20th century, the climate swings of the AMO have alternately camouflaged and exaggerated the effects of global warming, and have made attribution of global warming more difficult. Threshold of different phases of AMO signal is 0 and each phase lasts between 20 to 40 years.

Recent research has demonstrated that AMO has effects on the regional atmospheric circulation and on associated anomalies in precipitation and surface temperature over much of the Northern hemisphere, in particular, the United States, southern Mexico and probably Western

Europe and Sahel in Africa (Sutton and Hodson, 2005) as shown in Figure 8.Enfield et al. (2001) showed that there is a significant negative correlation with US continental rainfall, with less (more) rain over most of the central USA during a positive (negative) AMO index. According to Zhang and Delworth (2006) warm phase AMO strengthens the summer rainfall in India and Sahel in Africa and the Atlantic Hurricane activity. Apart from the research focusing on time mean anomalies, the index is also associated with changes in the frequency of extreme events. Droughts with broad impacts over the conterminous U.S. (1996, 1999-2002) were associated with North Atlantic warming (positive AMO) and northeastern and tropical Pacific cooling (negative PDO).When the AMO is in its warm phase, these droughts tend to be more frequent and/or severe and opposite occurs in the AMO negative phase (McCabe et al., 2004).

Monthly AMO index values comprising cold (negative) and warm phases (positive index) are accessible from the National Oceanic Atmospheric Administration (NOAA) site (http://www.esrl.noaa.gov/psd/data/timeseries/AMO/).The threshold that separates the phases is the zero line. As shown in Figure 9, warm phases occurred during 1860-1880 and 1930-1960, and cold phases occurred during 1905-1925 and 1970-1990. Since 1995, the AMO has been positive, and it seems the condition has persisted long enough to be considered a new warm phase.

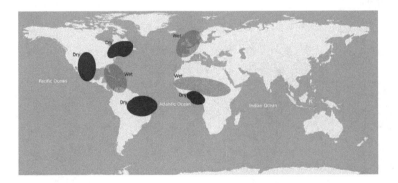

Figure 8. The regions having impacts of AMO

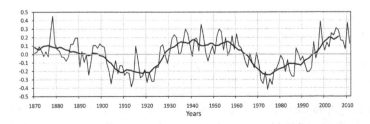

Figure 9. Fluctuation of the AMO index during 1870-2011 with a 10-year running average (bold line). The units on the vertical axes are °C.

6. Inter Tropical Convergence Zone (ITCZ)

ITCZ is the zone where the winds originating in the northern and southern hemispheres converge. In the Northern Hemisphere tropical region, the trade winds move in northeasterly direction, while in the Southern Hemisphere tropics, they are south easterly. Solar heating in the region results in a vertical motion in the air, leading to bands of clouds causing showers and thunderstorms around the equator. Unlike the zones north and south of the equator where the trade winds flow, within the ITCZ the average wind speeds are slight. Early sailors named this belt of calm *the doldrums* because of the inactivity and stagnation they found themselves in after days of no wind.

The location of the ITCZ varies throughout the year (Figure 10) and leads to different wet seasons in many equatorial regions. It also results in the wet and dry seasons in the tropics rather than the four seasons of higher latitudes. Over land, the north-south position follows a meandering path linked to the zenith of the sun, with width of several hundred kilometers, typically reaching a latitude of about 20° north or south, but with effects extending as far as 45° north in parts of southeast Asia. This often leads to two wet seasons and two dry seasons near the equator, merging into a single wet and dry season at the northern and southern limits (Sene, 2010).

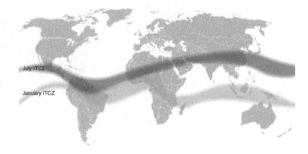

Figure 10. Movement of ITCZ during a year (source: Wikipedia, public domain)

7. Madden Julian Oscillation (MJO)

Scientists have identified many climatic phenomenon that fluctuate on time scales of several years to decades and multi decades, but they are now recognizing atmospheric anomalies on time scales of days to weeks that may influence the evolution of these longer oscillation processes. In 1971 Roland Madden and Paul Julian reported a 40-50 day oscillation when analyzing zonal wind anomalies in the tropical Pacific. This discovery was significant because it was believed until recently that the tropical weather variations on time scales less than one year are essentially random.

The MJO is an intra-seasonal atmospheric variability in the tropics. It is characterized by eastward propagation, moving slowly (~5 m/s) through the atmosphere and over the warm parts of the Indian Ocean into the west and central Pacific. It constantly interacts with the underlying ocean and influences many weather variations as well as lower and upper level wind speed and direction, cloudiness, rainfall, sea surface temperature and ocean surface evaporation (Madden and Julian, 1971; Zhang, 2005). There is a strong year-to-year variability in MJO activity, with periods of strong activity followed by long period in which the Oscillation is weak or absent (Zhang, 2005). In addition to strongly modulating the rainfall in the tropics and to a lesser extent in mid-latitudes; there is evidence that the MJO has effects on other meteorological variables like ENSO and the Indian monsoon. The inter-annual variability of the MJO is partly linked to the ENSO cycle. Strong MJO activity is often observed during weak La Niña years or during ENSO neutral years, while weak or absent MJO activity is typically associated with strong El Niño episodes (Kessler and Kleeman, 2000). Also the Australian, Asian, South American and North American monsoons can all be influenced by the MJO.

The climate prediction center of NOAA maintains an archive of MJO indices from 1978 to present which is accessible at the following address:

(http://www.cpc.ncep.noaa.gov/products/precip/CWlink/daily_mjo_index/mjo_index.html).

8. Sea Surface Temperature (SST) data

Ocean-atmospheric indices like PDO, NAO, AMO and SOI are commonly used as predictors in hydrological forecasting models but for some regions it makes sense to use the SSTs of nearby seas. Studying the coupled local SSTs and hydroclimatic events in regions such as United States (Tootle and Piechota, 2008), Colombia (Tootle et al., 2008), Brazil (Uvo et al., 1998) and Sri Lanka (Chandimala and Zubair, 2007) has assisted the identification of regions that are not represented in the existing series of Pacific, Atlantic and Indian Oceans SSTs. The mentioned studies resulted in greater correlation values than those of well-known climatic indices. Therefore, local experiments are highly recommended for evaluating effective regions of the seas in correlation with hydrological responses of local areas. For example for a country like Iran as well as several others in the Middle East two optimal candidates could be Mediterranean and Persian Gulf's sea surface temperatures. NOAA_OI_SST_V2 data provided by the NOAA/OAR/ESRL PSD, obtained from their Web site (http://www.esrl.noaa.gov/psd) are used as Mediterranean and Persian Gulf's SSTs. Reynolds et al. (2001) used *in situ* data observed from many ships, buoys and AVHRR satellite data, and utilized some modification methods. Reynolds Optimum Interpolation (OI.V2) monthly and weekly SST data are available from November 1981 to present for 1° grid squares of the seas.

8.1. Persian Gulf SST

The Persian Gulf, having a vast amount of crude oil and natural gas reserves of earth, is an important economic, military and political region. The shallowness of the Gulf (average depth ~36 m), the high evaporation rate, combined with a limited exchange through the Strait of

Hormuz cause the formation of a salty and dense water mass called Persian Gulf Water (PGW), which is apparently the warmest sea in the world with temperature reaching 35.6°C in the summer. The Persian Gulf is almost 990 km long extending from *Shatt-Al-Arab* (also called *Arvand-Roud*) as a major river source located at the head of the Gulf, which is fed by Euphrates, Tigris and Karoun rivers, to the Strait of *Hormuz* where it connects to the Oman Sea. It has a maximum width of 370 km with a surface area of about 239,000 km² (Emery, 1956). Extensive shallow regions (<20m) also occur along the coast of United-Arab-Emirates and Bahrain and Deep portions (>40m) are found along the Iranian coast continuing to the Oman sea (Kämpf and Sadrinasab, 2006).

The Strait of *Hormuz* acts as an exchange point between the Persian Gulf and Indian Ocean. The characteristic of the in/out flow at the strait has been investigated by many studies (e.g., Bower et al., 2000; John et al., 2003; Pous et al., 2004).At the head of the Oman Sea, where the Persian Gulf connects to Global Oceans, the fresher and colder inflow of Indian Ocean surface water (IOSW) core is clearly seen above and next to the Persian Gulf water outflow (Pous et al., 2004). In a research by John et al. (2003), the moored time series records show a relatively steady deep outflow through the strait from 40 m to the bottom with a mean speed of approximately 20 cm/s. A variable flow is found in the upper layer with frequent reversals on timescales of several days to weeks. The annual mean flow in the near-surface layer is found to be northeastward (out of the Persian Gulf) in the southern part of the strait, suggesting a mean horizontal exchange with the Indian Ocean that is superimposed on the vertical overturning exchange driven by evaporation over the gulf. Bower et al. (2000) studied the seasonal variability of the formation of dense PGW outflow and the surrounding oceanic environment with temperature and salinity profiles and found that the outflow salinity is slightly lower in winter (~0.5 practical salinity unit) than in summer and the temperature is cooler by about 3°C in winter.

Prevailing winds over the Persian Gulf are northwesterly. During winter (November-February) the winds are slightly stronger (~5 m/s) than those during the summer (June-September) (~3 m/s). The best known phenomenon in the Persian Gulf that can cause abrupt changes in the circulation and heat-budget for a short period of time is *shamal*, a northwesterly wind occurring during winter as well as summer (Perrone, 1979). Based on duration, there are two types of winter *shamal*: those which last 24-36 hours and those which last for a longer period of 3 to 5 days. The winds bring cold dry air (T=20°C) and result in SST cooling of about 10°C noticeably in the northern and shallower shelf regions (Thoppil and Hogan, 2010). The other air masses affecting the area, mostly in winter, is Sudan current. About 30% of the total rain-bearing air masses coming to Iran originate in North Africa, Red Sea and Saudi Arabia which is called Sudan current (Khalili, 1992) (Figure11).

The general water circulation of the Persian Gulf is cyclonic, which is bounded by an Iranian Coastal Current (ICC) flowing northwestward along the northern side from the Strait of *Hormuz* with speed greater than 10cm/s, and a southeastward current in the southern part of the Gulf. The ICC, which flows against the prevailing northwesterly winds, is primarily driven by the pressure gradient. The dense bottom outflow forming in the southern shallows flows along the coastline of UAE (Hunter, 1983; Reynolds, 1993).

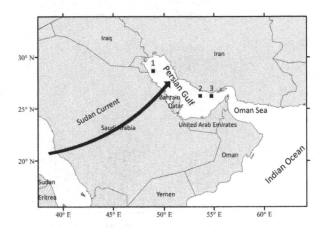

Figure 11. Persian Gulf location and general track of air masses toward Iran during winter

The Persian Gulf SSTs could be used as possible predictors of hydrological parameters for countries around it, including Iran. It has been shown (Nazemosadat, 1998)that winter droughts and wet periods of South western Iran tend to coincide with periods when the Persian Gulf SSTs are above (below) normal. Overall, correlation analysis between rainfall and SST data, using various data lengths, has revealed that the fluctuations of SST account for about 40% of rainfall variability over the region.

Using OI.V2 data, the anomalies of monthly SST values for three selected points shown in Figure11 are demonstrated as an example in Figure 12. It is shown that the eastern parts are possibly warmer than northwestern ones. Besides, a warming trend is noted for all points by applying a trend line for the annual mean values (dotted line). For example, the mean annual warming rate for point 2 is calculated at 0.035°C/year which should be considered in long-term forecasting. A noteworthy point, that needs more analysis, is the similar fluctuation of annual values in all three diagrams showing a possible change phase every 2 to 4 year.

8.2. Mediterranean SST

The Mediterranean Sea is located in the middle of the Mediterranean. The main rainy season over the Mediterranean Sea extends from October to March, but maximum rainfall occurs during November-December. During the rainy season western Mediterranean Sea receives ~20% larger rainfall compared to eastern Mediterranean Sea (Mehta and Yang, 2008). Mediterranean Sea has climatic dynamics that affect the climate of various areas around, including Sahel in Africa (Rowell, 2003), Turkey (Kutiel, 2001), Iran (Araghinejad and Meidani, 2012), and Greece (Kassomenos and McGregor, 2006). Due to its size and limited exchange at Gibraltar Strait, Mediterranean Sea dynamics is mainly linked to the local climate and it is particularly sensitive to anthropogenic climatic perturbations (Skiliris et al., 2011). Several studies (e.g. Rixen, 2005; Belkin, 2009) have demonstrated the rapid surface warming of the sea during the last decades that should be considered in long-lead forecasting when these SSTs

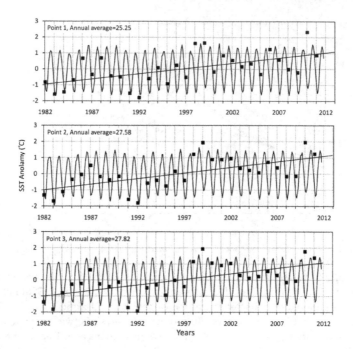

Figure 12. Fluctuations of Persian Gulf SSTs for selected points during 1982-2011.

are used as predictors. Skilirish (2011) reported the satellite-derived mean annual warming rate of about 0.037°C/year for the whole basin, about 0.026°C/year for the western sub-basin and about 0.042°C/year for the eastern sub-basin over 1985-2008.

During the northern hemisphere winter, the high pressure cells over the north Atlantic are centered at about 30°N and those over Asia at about 45°N. Between them lies the Mediterranean region, an area of warm and moist air, with a tendency for low pressure. So along the north coast of Africa all the way from Morocco to Egypt, the prevailing winds in winter are westerly, controlled by the low pressure over the Mediterranean. During the northern summer the subtropical high pressure cells of the northern hemisphere are found in well-marked anti cyclone over the oceans. The Asia and Africa continents are relatively hot, the air over them expands, and a low pressure system results, changing the prevailing winds direction from north to south and southwest (Kendrew 1922). As shown in Figure 13, the Mediterranean currents influence mostly west and, to a lesser extent, the center of Iran in winter.

Rezaebanafsheh (2011) studied the relationship between winter and autumn precipitation with previous season anomalies of Mediterranean SST at several stations in western Iran and showed that the cooler condition of the sea in autumn leads to wetter winter in the study area, but the correlation of the summer SST anomalies with autumn precipitation was not significant.

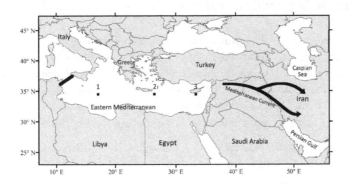

Figure 13. Mediterranean Sea location and general track of air masses toward Iran during winter

Rainfall amounts differ markedly from place to place over the Mediterranean. Sea surface temperatures are highest in the eastern Mediterranean compared to elsewhere in the sea. Fluctuation of standard values of average monthly SSTs over the Mediterranean Sea, using OI.V2 data are shown in Figure 14. Apart from the warming rate mentioned in the literature, similar to Persian Gulf, the annual values of the SSTs (dotted line) show a shift in phase every 2 to 4 year.

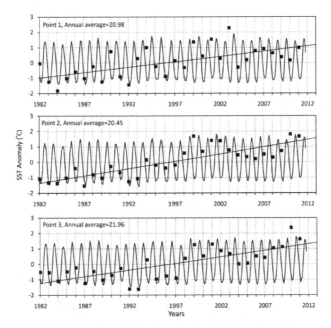

Figure 14. Fluctuations of Mediterranean SSTs for selected points during 1982-2011

9. The effects of large scale climate signals around the world

The effects of large scale climate signals on different locations of the world need to be studied and investigated in order to identify the specific dominant signals that impact each region. As mentioned previously, such background information provides useful context for assessing pattern changes due to climate change. The results could also be used for long-lead hydroclimatic forecasts. Tables 2 to 5 are selected compilation of various teleconnections and their hydro-climatological impacts around the world as documented in easily accessible literature.

The table shows researchers, the location of the study, the preferred climate signals as predictors, an overview of the paper as well as the time period that the study was conducted.

Reference	Location	Time period*	Preferred Climatic predictors	Overview
Ropelewski and Halpert, 1987	Eastern Africa	Various time periods	ENSO	During the LaNina event, the eastern Ecuadorian Africa experiences more than normal rainfall meanwhile the eastern sought of Africa experiences less than normal rainfall
Nicholson and Kim, 1997	Africa	1901-1990	ENSO	La Nina and El Nino cause wet and dry years, respectively. However the type of impacts of El Nino has a more variation in different time and locations
Nicholson and Selato, 2000	Africa	1901-1990	ENSO	They obtained the same results as Ropelewski and Halpert, (1987) by studying a wider region
Rowell, 2003	Sahel in Africa	1947-1996	Mediterranean SSTs	Warmer than average Mediterranean leads to a wetter than normal Sahel

* Time period covered in the study

Table 2. A select list of studies showing locations in Africa where teleconnection indices are related to hydrological parameters

Reference	Location	Time period	Preferred Climatic predictors	Overview
Schonher and Nicholson, 1989	California	1950-1980	ENSO	There is a relationship between ENSO and wet years
Mantua et al., 1997	North America	1900-1996	PDO	During warm PDO years, wintertime precipitation decreases over much of the interior of North America and Alaskan salmon catches increases
Uvo et al., 1998	Northeast Brazil	1946-1985	Pacific and Atlantic SSTs	During April-May, North tropical (neg) and South tropical Atlantics (pos) and equatorial pacific (neg) are most correlated regions

Reference	Location	Time period	Preferred Climatic predictors	Overview
Jones, 2000	The U.S. (California state)	1958-1996	MJO	Slightly more extreme rainfall occur as convective anomalies are located over the Indian Ocean
Enfield et al., 2001	United States	1856-1999	AMO	AMO warmings decrease annual rainfall over U.S., especially over the eastern Mississippi basin
Karla and Ahmad, 2009	Western United States	1906-2001	ENSO and NAO	ENSO and NAO are better predictors for long-lead Streamflow forecasting of Upper Colorado River Basin as compared to PDO and AMO
Kayano and Sansigolo, 2009	Southern Brazil	1913-2006	ENSO (Nino 3.4), SSTs of southwestern subtropical Atlantic	Excessive rainfall during El-Niño and higher TMIN by warming of SSA ocean surface water
Barrett et al., 2012	Chile	1980-2010	MJO	Positive precipitation anomalies in central and south-central Chile for MJO phases 8, 1 and 2, and negative anomalies in phases 3-7
Tootle and Piechota, 2006	United States	1951-2002	North central and near the tropical Pacific, northern South American coast and northern Atlantic Ocean are most correlated regions, considering all years	PDO, ENSO and AMO were acknowledged as effective regions on the continental US streamflow variability
Tootle et al., 2008	Colombia	1960-2000	Eastern coast of Australia, Equatorial and South central Pacific SSTs as most correlated regions	An El-Niño (La-Niña) will result in decreased (increased) streamflow
Karla and Ahmad, 2009	Western United States	1906-2001	ENSO and NAO	ENSO and NAO are better predictors for long-lead Streamflow forecasting of Upper Colorado River Basin as compared to PDO and AMO
Kayano and Sansigolo, 2009	Southern Brazil	1913-2006	ENSO (Nino 3.4), SSTs of southwestern subtropical Atlantic	Excessive rainfall during El-Niño and higher TMIN by warming of SSA ocean surface water
Barrett et al., 2012	Chile	1980-2010	MJO	Positive precipitation anomalies in central and south-central Chile for MJO phases 8, 1

Reference	Location	Time period	Preferred Climatic predictors	Overview
				and 2, and negative anomalies in phases 3-7

* Time period covered in the study

Table 3. A select list of studies showing locations in America where teleconnection indices are related to hydrological parameters

Reference	Location	Time period	Preferred Climatic predictors	Overview
Nazemosadat, 1998	Southern Iran	1951-1987	Persian Gulf SSTs	Winter droughts and wet periods tend to occur when PGSSTs are above and below normal, respectively
Kutiel et al., 2001	Eastern Mediterranean region over Turkey	1929-1996	Sea Level Pressure (SLP)	Positive SLP departures are associated with dry conditions of Turkey and vice versa. The relationship is large in winter
Wheeler and Hendon, 2004	Australia and India	1979-2001	MJO	Monsoons occur mostly during the enhanced half of the MJO cycle and rarely during the suppressed half. In Darwin, Australia, the likelihood of extreme rainfall becomes triple from dry to wet phase of MJO
Power et al., 2005	Australia	Over 100 years	ENSO	In general, El-Niño events are associated with drying and la-Niña events with increased rainfall, but, asymmetrically.
Araghinejad et al., 2006	Iran, Isfahan	1969-2001	SOI, NAO and Snow Budget	Higher SOI values (Jan-Oct) corresponds to drier autumn and winter seasons and higher NAO values corresponds to wetter spring
Sen Roy, 2006	India and eastern Arabian Sea	1925-1998	ENSO and PDO	There is negative relationship between Winter precipitation and both ENSO and PDO indices
Chandimala and Zubair, 2007	Sri Lanka	1950-2000	ENSO, and Indian Ocean SSTs	Reduced streamflow (Apr-Sep) and enhanced rainfall (Oct-Dec) during El-Niño are dominant
Zhang et al., 2009	Southeast China	1980-2010	MJO	The perceptible water decreases on intraseasonal time scales as the

Reference	Location	Time period	Preferred Climatic predictors	Overview
				convective center moves from the Indian Ocean to the western Pacific Ocean
Zhu et al., 2010	East China	1951-2008	PDO	The shift of the PDO to a negative phase increases rainfall in the Huang-Huai river region and decreases in the Yangtze river region

* Time period covered in the study

Table 4. A select list of studies showing locations in Asia and Australia where teleconnection indices are related to hydrological parameters

Reference	Location	Time period	Preferred Climatic predictors	Overview
Feudale and Shukla, 2007	Europe	2003	Mediterranean SSTs	The SSTs are able to simulate the upper level anticyclone over central Europe, but to a lesser extent than global SST.
Hurrel et al., 2003	North America, North Africa, Europe and Middle East	-	NAO	During NAO positive phase, much of central and southern Europe, Mediterranean and Middle East are drier, eastern U.S. and Iceland through Scandinavia are wetter, North Africa and Middle East are cooler and North America is warmer
Sutton and Hodson, 2005	The U.S., Europe and Sahel in Africa	1931-1990	AMO	AMO warm phase cause decreased summer precipitation and warmer temperature in the U.S., increased precipitation in Europe and Sahel and warmer temperature in western Europe
Zhang and Delworth, 2006	India and Sahel in Africa	1901-2000	AMO	AMO, with a positive correlation, play a leading role in the 20th century multidecadal variation of India/Sahel summer rainfall and Atlantic Hurricane activity
Brandimarte et al., 2011	Mediterranean Areas (southern Italy and Nile Delta in Egypt)	1920-1996	NAO	Hydroclimatic variables of most of the Mediterranean areas are negatively correlated with NAO, while southeastern parts have a weaker positive correlation

* Time period covered in the study

Table 5. A select list of studies showing locations in Europe and multi-continents where teleconnection indices are related to hydrological parameters

10. Summary

The chapter reviewed the status of knowledge concerning teleconnection ocean-atmospheric dynamics. Different climate signals such as Southern Oscillation (SO), Pacific Decadal Oscillation (PDO), North Atlantic oscillation (NAO), Atlantic Multi-decadal Oscillation index (AMO), Madden Julian Oscillation (MJO), Inter Tropical Convergence Zone (ITCZ), and Sea Surface Temperature (SSTs) of two large scale water bodies were reviewed. In addition to providing basic information, the reports on the effect of those signals in different regions of the world were reviewed in terms of their potential as long-lead hydroclimatological predictors of rainfall. The effects of climatic signals on different regions were reviewed based on the recent studies of different researchers and scientists. The information provided is useful for assessing probable changes in the frequency and magnitude of these teleconnection indices as a result of climate change.

Author details

Shahab Araghinejad and Ehsan Meidani

College of Agricultural Technology and Science, University of Tehran, Karaj, Iran

References

[1] Araghinejad, S., Burn, D.H., and Karamouz, M. (2006). "Long-lead probabilistic forecasting of streamflow using ocean-atmospheric and hydrological predictors." Water Resour. Res., Vol. 42, DOI: 10.1029/2004WR003853.

[2] Araghinejad, S., Meidani, E. (2012). "Probabilistic Drought Forecasting for a Basin Scale Water Resources Operation." HydroPredict2012 conference, Vienna, Austria, 24-27 Sep.

[3] Barrett, B., Carrasco, J.F., Testino, A.P. (2012). "Madden–Julian Oscillation (MJO) Modulation of Atmospheric Circulation and Chilean Winter Precipitation." J. Clim., Vol. 25, pp 1678-1688. DOI:10.1175/JCLI-D-11-00216.1

[4] Beebee, R.A., and Manga, M. (2004). "Variation in the relationship between snowmelt runoff in Oregon and ENSO and PDO." J. Am. Water Resour. Assoc., 40(4), 1011–1024, DOI:10.1111/j.1752-1688.2004.tb01063.x.

[5] Belkin, M. (2009). "Rapid warming of large marine ecosystems." Progr. Oceanogr., 81, 207–213.

[6] Biondi, F., Gershunov A., and Cayan D.R., (2001): North Pacific decadal climate variability since 1661. J. Climate, 14, 5–10.

[7]   Bower, A.S., Hunt H.D., and Price J.F. (2000). "Character and dynamics of the Red Sea and Persian Gulf outflows." J. Geophys. Res., 105(C3), pp 6387-6414.

[8]   Brandimarte, L., Baldassarre, G.D., Bruni, G., D'Odorico, P., Montanari, A., (2011). "Relation between the North-Atlantic Oscillation and Hydroclimatic Conditions in Mediterranean Areas." Water Resour Manage, Vol. 25, pp 1269–1279. DOI 10.1007/s11269-010-9742-5.

[9]   Chandimala, J., Zubair, L. (2007). "Predictability of stream flow and rainfall based on ENSO for water resources management in Sri Lanka." J. Hydrol. Vol. 335, pp 303–312

[10] Collins, M., and Sinha, B. (2003). "Predictability of decadal variations in the thermohaline circulation and climate." Geophy. Rese. Let. Vol. 30, NO. 6, 1306, DOI: 10.1029/2002GL016504

[11] Dettinger, M.D., and Diaz, H.F. (2000). "Global Characteristics of Streamflow Seasonality and Variability." J. Hydrometeorology, Vol. 1, pp 289-310

[12] Enfield, D.B., Mestas-Nunez, A.M., and Trimble, P.J. (2001). "The Atlantic Multidecadal Oscillation and its relation to rainfall and river flows in the continental U.S." Geophy. Rese. Let., Vol. 28, No.10, pp 2077-2080.

[13] Emery, K O.1956. "Sedimentsand and waters of Persian Gulf", Bull. Amer. Assoc.

[14] Feudale, L., and Shukla, J., (2007). "Role of Mediterranean SST in enhancing the European heat wave of summer 2003." Geophy. Rese. Let., Vol. 34, DOI: 10.1029/2006GL027991.

[15] Gedalof, Z., and Smith D.J., (2001): Interdecadal climate variability and regime-scale shifts in Pacific North America. Geophys. Res. Lett., 28, 1515–1518.

[16] Gray, S.T., Graumlich, J.L., and Pederson, G.T. (2004). "Atree-ring based reconstruction of the Atlantic Multidecadal Oscillation since 1567 A.D." Geophys. Res. Lett., 31, DOI:10.1029/ 2004GL019932.

[17] Hunter, J.R. (1983). "Aspects of the dynamics of the residual circulation of the Arabian Gulf, in: Coastal oceanography." edited by: Gade, H. G., Edwards, A., and Svendsen, H., Plenum Press, 31– 42.

[18] Hurrell, J.W. (1995). "Decadal trends in the North Atlantic Oscillation: Regional temperatures and precipitation." Science, 269(5224), pp 676– 679, DOI:10.1126/science.269.5224.676.

[19] Hurrell, J.W., and vanLoon, H., (1995). "Decadal Variation in climate associated with the North Atlantic Oscillation." Climate Change, No. 31, pp 301-326.

[20] Hurrell, J.W., and Deser, C. (2009). "North Atlantic climate variability: The role of the North Atlantic Oscillation." J. Mar. Syst., Vol. 78, No. 1, pp 28-41

[21] Hurrell, J.W., Kushnir, Y., Ottersen, G. (2003). "An overview of the North Atlantic oscillation." In: Hurrell, J.W, Kushnir, Y., Ottersen, G., Visbeck, M., (eds) The North Atlantic Oscillation: Climatic Significance and Environmental Impact. American Geophysical Union, pp 1-35.

[22] Johns W.E., F. Yao, D.B. Olson, S.A. Josey, J.P. Grist and D.A. Smeed, 2003, "Observations of seasonal exchange through the Straits of Hormuz and the inferred heat and freshwater budgets of the Persian Gulf." J. Geophys. Res., 108, C12, 3391, doi: 10.1029/2003JC001881.

[23] Jones, C. (2000). "Occurrence of Extreme Precipitation Events in California and Relationships with the Madden–Julian Oscillation." J. Clim., Vol. 13, pp 3576-3587.

[24] Jones, P.D., Jonsson, T., Wheeler, D. (1997). "Extension to the North Atlantic Oscillation using early instrumental pressure observations from Gibraltar and south-west Iceland." Int. J. Climatol. Vol. 17, pp 1433–1450.

[25] Kämpf, J., Sadrinasab, M. (2006). "The circulation of the Persian Gulf: a numerical study." Ocean Science Disc., Vol. 2, pp 1-15.

[26] Karla, A., Ahmad, S., (2009). "Using ocean atmospheric oscillation for long lead time streamflowforcasting". WaterResour. Res., Vol. 45, DOI:10.1029/2008WR006855.

[27] Kassomenos, P.A., and McGregor, G.R. (2006). "The Interannual Variability and Trend of Precipitable Water over Southern Greece." J. Hydromet., Vol. 7, pp 271-284.

[28] Kayano, M.T., Sansígolo, C. (2009). "Interannual to decadal variations of precipitation and daily maximum and daily minimum temperatures in southern Brazil." Theor. Appl. Climatol., Vol. 97, pp 81–90, DOI:10.1007/s00704-008-0050-4

[29] Kendrew, W.G. (1922). "The Climates of the Continents." Oxford University Press.

[30] Kessler, W., and Kleeman, R., (2000). "Rectification of the Madden-Julian Oscillation into the ENSO." cycle. J. Climate, Vol. 13, pp 3560-3575.

[31] Khalili, A. (1992). "Fundamental Study of Iranian Water Resources, Climatological Understanding of Iran, Parts 1 and 2." Jamab consultant reports, the Iranian Ministry of Energy (in Persian).

[32] Kutiel, H., Hirsch-Eshkol, T. R., and Turkes, M. (2001). "Sea level pressure patterns associated with dry or wet monthly rainfall conditions in Turkey." Theor. Appl. Climatol., Vol. 69, pp 39-67.

[33] Madden, R.A., and Julian, P.R. (1971). "Detection of a 40-50-day oscillation in the zonal wind in the tropical Pacific." J. Atmos. Sci., Vol. 28, pp 702–708.

[34] Mantua, N. J., and Hare, S. R. (2002). "The Pacific decadal oscillation." J. Oceanogr., Vol. 58, pp 35–44.

[35] Mantua, N.J., Hare, S.R., Zhang, Y., Wallace, J.M., Francis, R.C. (1997). "A Pacific interdecadal climate oscillation with impacts on salmon production." Bulletin of the American Meteorological Society, Vol. 78, pp 1069-1079.

[36] McCabe, G.J., Palecki, M.A., Betancourt, J.L. (2004). "Pacific and Atlantic Ocean influences on multidecadal drought frequency in the United States." Nat. Acad. Sci. U.S.A., volume 101, 4136-4141

[37] Mehta A.V. and Yang, S., 2008: Precipitation Climatology over Mediterranean Basin from TRMM, Advances in Geosciences, 17, 87-91.

[38] Nazemosadat, M.J. (1998). "The Persian Gulf sea surface temperature as a drought diagnostic for southern parts of Iran." Drought News Network, Vol. 10, pp 12-14.

[39] Nicholson S. E. and Kim J., "The relationship of the El Nino-Southern Oscillation to African rainfall" International Journal of Climatology. Vol. 17, 1997. pp. 117 – 135

[40] Nicholson, S. E. and Selato, J. C., "The influence of LaNina on African rainfall", Int. J. of Climatology, Vol. 20, 2000. pp. 1761-1776

[41] Perrone, T.J. (1979). "Winter shamal in the Persian Gulf." Naval Env. Prediction Res. Facility. Technical Report. 79-06. Monterey, 180pp.

[42] Portis, D.H., Walsh, J.E., El Hamly, M., Lamb, P.J. (2001). "Seasonality of the North Atlantic Oscillation." J. Climate., Vol. 14, pp 2069–2078.

[43] Pous, S.P., Carton, X., Lazure, P. (2004). "Hydrology and circulation in the Strait of Hormuz and the Gulf of Oman—Results from the GOGP99 Experiment: 1. Strait of Hormuz." J. Geophy. Rese., Vol. 109; PART 12; SECT 3; pp. C12037.

[44] Power, S., Haylock, M., Colman, R., Wang, X., 2005. Asymmetry in the Australian response to ENSO and the predictability of interdecadal change sin ENSO teleconnections. Bureau of Meteorology Research Centre. BMRC Report No. 113, 33pp.

[45] Reynolds, R.M. (1993). "Physical oceanography of the Gulf, Strait of Hormuz, and the Gulf of Oman – Results from the Mt Mitchell expedition." Mar. Pollution Bull., Vol. 27, pp 35–59.

[46] Reynolds, R.W., Rayner, N.A., Smith, T.M., Stokes, D.C., and Wang, W. (2001). "An Improved In Situ and Satellite SST Analysis for Climate." Science Applications International Corporaton, Camp Spring, Maryland.

[47] Rezaebanafsheh, M., Jahanbakhsh, S., Bayati, M. and Zeynali, B. (2011). "Forecasting autumn and winter precipitation of west of Iran applying Mediterranean SSTs in summer and autumn." Phy. Geog. Research Quarterly, Vol. 74, pp 47-62.

[48] Rixen, M. (2005). "The Western Mediterranean deep water: a proxy for climate change." Geophys Res. Lett., DOI:10.1029/2005GL022702.

[49] Ropelewski, C.F. and M.S. Halpert, "Global and regional scale precipitation patterns associated with the El Niño/Southern Oscillation". Mon. Wea. Rev., Vol. 115, 1987. pp. 1606-1626.

[50] Rowell, D.P. (2003). "The Impact of Mediterranean SSTs on the Sahelian Rainfall Seasonal." J. Clim., Vol. 16, 5, 849–862.

[51] Schonher, T., and S. E., Nicholson, "The relationship between California rainfall and ENSO events", J of Climate, Vol. 2, 1989. pp. 1258-1269

[52] Sene, K. (2010). "Hydro-meteorology forecasting and applications." Springer, DOI: 10.1007/978-90-481-3403-8

[53] Sen Roy, S., (2006). "The Impacts of ENSO, PDO and local SSTs on winter precipitation in India." Physical Geography, Vol. 27, 5, pp 464-474.

[54] Skiliris, N., Sofianos, S., Gkanasos, A., Mantiziafo, A., Vervatis, V., Axaopoulos, P., and Lascaratos, A. (2011). "Decadal scale variability of sea surface temperature in the Mediterranean Sea in relation to atmospheric vriability." Ocean Dyn., DOI: 10.1007/s10236-011-0493-5.

[55] Sutton, R.T., and Hodson, D.L.R. (2005). "Atlantic Ocean forcing of North American and European summer climate." Science, 309, 115–118.

[56] Thoppil, P.G., Hogan, P.J. (2010). "Persian Gulf response to a wintertime shamal wind event." Deep-Sea Research, 1, 57, pp 946-955, DOI: 10.1016/j.dsr.2010.03.002.

[57] Tootle, G.A., Piechota, T.C. (2006). "Relationships between Pacific and Atlantic ocean sea surface temperatures and US streamflow variability." Water Resources Research, 42, W07411.

[58] Tootle, G.A., Piechota, T.C., Gutierrez, F. (2008). "The relationships between pacific and Atlantic ocean sea surface temperatures and Colombian streamflow variability." J. Hydrol., 349, 268–276.

[59] Uvo, C.B., Repelli, C.A., Zebiak, S.E. and Kushnir, Y. (1998). "The relationships between tropical Pacific and Atlantic SST and northeast Brazil monthly precipitation." J. Clim., Vol. 11, pp 551–562.

[60] Wheeler, M.C., and Hendon, H.H. (2004). "An All-Season Real-Time Multivariate MJO Index: Development of an Index for Monitoring and Prediction." Monthly Weather Review, Vol. 132, pp 1917-1932.

[61] Zhang, C., (2005). "Madden-Julian Oscillation." Rev. Geophys., Vol. 43, DOI: 10.1029/2004RG000158.

[62] Zhang, R., Delworth, T.L. (2006). "Impact of Atlantic multidecadal oscillations on India/Sahel rainfall and Atlantic hurricanes." Geoph. Res. Let., Vol. 33, DOI: 10.1029/2006GL026267.

[63] Zhang, Y., Wallace J.M., Battisti, D.S (1997). "ENSO-like interdecadal variability: 1900-93." J. Clim., Vol. 10, pp 1004-1020.

[64] Zhang, l., Wang, B., Zeng, Q.(2009). "Impact of the Madden–Julian Oscillation on Summer Rainfall in Southeast China." J. Clim., Vol. 22, pp 201-216, DOI: 10.1175/2008JCLI1959.1

[65] Zhu, Y., Wang, H., Zhou, W., Ma, J. (2010). "Recent changes in the summer precipitation pattern in East China and the background circulation." Clim. Dyn., DOI: 10.1007/s00382-010-0852-9.

Chapter 7

High-Resolution Surface Observations for Climate Monitoring

Renee A. McPherson

Additional information is available at the end of the chapter

1. Introduction

Climate monitoring is a foundational component of understanding climate variability across the globe and regionally. In the United States and other countries, surface measurements of temperature and precipitation have been taken routinely by human observers (longest records) and by automated stations overseen by the federal government. Budget plateaus or reductions have resulted in the elimination of many observing stations as well as delays in adopting new measurement technologies or techniques. The future of climate monitoring may look bleak at a time when knowledge of our climate record is critical for water resource management, military and emergency preparedness, infrastructure planning, agricultural production, tourism, and science education.

Fortunately, innovation has accelerated in the academic and private sectors, leading to new climate monitoring infrastructure that is revitalizing how the nation might address these growing challenges. The establishment of regional "mesonets" — surface observing networks that measure atmospheric and soil variables at least hourly at spatial resolutions of 10-50 km — has not only been successful in providing meteorologists with vital, real-time information for weather forecasting but has helped to document previously unknown or seemingly rare regional climate phenomena. These networks also provide high-resolution data for increased understanding of climate variability in both space and time. Data from high-quality, real-time mesonets now are used extensively by researchers, educators, and practitioners. It appears that these networks will become a vital component of our long-term monitoring capacity in the United States.

This chapter will discuss the current status of monitoring the climate of the United States using surface observing stations, the evolution of regional mesonets, examples of new knowledge generated about regional climate variability as a result of these mesonets, and what the future may hold for high-resolution climate monitoring through a "network of networks." The proposed "network of networks" brings its own challenges to monitoring the climate, as many of the advocates of high-resolution, real-time monitoring are those interested in observations for weather forecasting only. Hence, the required standards for obtaining climate-quality data may be considered too costly for state and private network operators. This chapter will provide rationale for still pursuing this course, as well as continuing our federal observation program.

2. Brief history of surface climate observations

Since Wren's invention of the tipping bucket rain gauge in 1662 and Fahrenheit's invention of the mercury thermometer in 1714, systematic measurements of the weather have been taken at individual locations across various countries. In Europe, for example, monthly precipitation observations recorded at the Paris Observatory (France) began in 1688 [1] and monthly mean temperature records for central England have been constructed from several locations and time series since 1698 [2]. In the United States, instrumented observations began in Cambridge, Massachusetts; Boston, Massachusetts; and Philadelphia, Pennsylvania in the early 1700s [3]. These and other weather observations required meticulous and disciplined manual measurements and recording, typically conducted by scientists at national institutions or observatories.

At the First International Meteorological Congress in Vienna, Austria, during September 1873, discussions by the existing national meteorological services firmly instituted the development of an international (albeit spatially inhomogeneous) network of standardized weather observations [4]. Technical work at this Congress included the standardization of the time of observation, observational methods and units, and instrument testing and calibration. In addition, the participating countries agreed to share their data with one another in a timely fashion (i.e., via telegraph). With standardization and documentation, a true surface climate observation network could be advanced worldwide, and national meteorological services became increasingly important for strengthening their nation's observational capabilities.

A century would pass, however, before automation of surface weather measurements became inexpensive enough for deployment in most nations. Until then, daily temperature and precipitation — the heart of our climate record — were measured and recorded manually by both experienced observers and trained volunteers. Experienced observers at U.S. Weather Bureau/National Weather Service offices reported an array of weather variables every hour, including atmospheric pressure, temperature, wind speed and direction, humidity, precipitation type and amount, cloud cover, and visibility. In the United States, because of its vast size and limited funding, an extensive network of volunteer "cooperative observers" also was needed to capture the spatial and temporal variability of the nation's climate [3]. The Organic Act of 1890, which also charged the Weather Bureau with recording meteorological observations, created the U.S. Cooperative Observer Program, thus forming the backbone of the modern climate record of the United States. At its peak in the late 1950s, this network had about 14,000 observers nationwide. Today's Cooperative Observer Program has a two-fold mission: (1) to provide meteorological observations necessary to define the climate of the

United States and to document changes in the climate; and (2) to provide meteorological observations in near real-time for forecast, warning, and other public service operations of the National Weather Service (formerly the Weather Bureau) [5]. Cooperative observer data usually include daily maximum and minimum temperatures, snowfall, and daily precipitation totals.

During the period from the international recognition of a need for standardization and data sharing until the advent of reliable and affordable automated weather stations, a distinction between synoptic weather observations and climatological measurements became apparent and, in some nations, engrained in the culture of the government. As noted by Eden (2009) for this division within the United Kingdom:

By 1880, the contrast between synoptic and climatological observations, which first manifested itself some three decades earlier, had become much more clear-cut. Synoptic observations of pressure, temperature, wind, sky state, present weather, and visibility, were taken at specific times of the day and immediately transmitted by electric telegraph to central offices where they were plotted on maps which then had isobars and other features drawn on them. Climatological observations were generally taken once per day, normally at 0900, and included non-synoptic elements such as maximum and minimum temperature and rainfall for the preceding 24 hours, run of the wind, state of the ground, soil temperatures, and sunshine duration (starting in 1880 at a selection of sites); they often included a manuscript weather diary as well. They were tabulated and sent by post to a central office either weekly or monthly. [6]

Thus, the onset of a national Automated Surface Observing System (ASOS) by the U.S. National Weather Service, as part of the organization's $4.5 billion modernization from 1989 to 2000 [7], was seen by some as advancing weather (synoptic) observations to the detriment of the climate record. Part of the concern by the climatological community was the deployment of ASOS and other automated weather stations at airports, where the observations may not be representative of the region's climate. Another concern was the large investment in ASOS funding likely was to result in reduced instrumentation equipment, maintenance, and operations funding for the cooperative observer program, thus jeopardizing the quantity and quality of the climate records. By the beginning of the 21st century, the number of cooperative observing sites had, in fact, nearly halved from its mid-20th century peak [3].

At the same time, there was substantial growth in automated weather observing outside of the federal sector. In particular, the private sector moved quickly into developing observing programs and, with several examples of companies that cut corners on system deployment or operation, the climate community again declared concern for the future of the climate record. Over the years, network operators began to realize the benefit of increasing the quality of their sensors, methods, and data management. In part, this change resulted from recommendations by the National Research Council (NRC 1999) that network operators adopt the following 10 climate monitoring principles, proposed by Karl et al. (1995; [8]):

1. Management of network change: Network operators should critically examine how changes in their network (e.g., station location, data processing, instrumentation) may impact climatological analyses of the locale and its region.

2. Parallel testing: When changes in the network do occur, network operators should develop appropriate transfer functions for time series by operating the old and new systems (locations, instruments, etc.) simultaneously over a sufficiently long time period to observe the range of climate variability.

3. Metadata: Network operators should fully document the observing system and its operating procedures, including station location and exposure, instrumentation, sampling times, calibration and validation methods, quality assurance procedures, data processing algorithms, and other information pertinent to the data history.

4. Data quality and continuity: Network operators should routinely assess quality and homogeneity of all observations.

5. Integrated environmental assessment: Network operators should plan for their data to be used in state, regional, national, or international climate assessments. Regular scientific analysis of data time series adds value to the monitoring program.

6. Historical significance: Network operators should identify protected sites within their network that will be maintained for many decades to a century or more. Sites should be prioritized based on their contribution to obtaining a homogeneous, long-term climate record.

7. Complementary data: Network operators should prioritize funding for data-poor regions, poorly observed variables, regions that are sensitive to land use/land cover or other changes, and high temporal resolution for critical measurements.

8. Climate requirements: Network designers, operators, and instrument engineers should be provided climate monitoring requirements appropriate to the region and consistent with national and international standards at the outset of network design.

9. Continuity of purpose: Network operators should commit to long-term and stable observations while also serving short-term operational needs.

10. Data and metadata access: Network operators should develop data management systems that enable users to easily and cost-effectively access, use, and interpret data and data products.

These or similar principles inspire network managers to care for their observations prior to distribution to thousands of data users. For the government, in particular, additional operational costs needed for data quality assurance, maintenance, and data and metadata management greatly reduce the expense (multiplied by hundreds of thousands to millions of users) for quality control procedures on the client side. Hence, budget officials must become aware of the net savings to the economy when data are cared for properly by the network owner.

3. Current surface observation capabilities

The surface observational networks of the United States provide excellent examples of the capabilities and challenges for monitoring climate variability and change across diverse land-scapes and providing data to diverse users. The networks can be generally characterized as the following: (1) climate reference networks, (2) government-operated volunteer observer networks, (3) mesoscale weather and climate monitoring networks, (4) use-inspired surface observing networks, and (5) citizen science networks. Climate reference networks are designed, developed, deployed, operated, and maintained specifically to obtain a high-quality climate record that is representative of a region. Government-operated volunteer observer networks are designed to obtain daily observations with high spatial resolution for a long-term climate record. Mesoscale weather and climate monitoring networks are multi-purpose networks that strive for high-quality observations that can be applied in near real-time to critical and non-critical decisions across multiple disciplines. Use-inspired surface observing networks are designed to solve a specific problem or gap in the current observing capabilities. Citizen science networks help to fill significant gaps in the spatial coverage of other networks as well as engaging and educating the public about the importance of data in scientific research and decision making. Certainly any given observing system or network may fit into multiple categories, but most public- and private-funded networks fall primarily into one of these network types.

3.1. Climate reference networks

The premier climate reference network is the U.S. Climate Reference Network (USCRN), which first deployed operational stations in 2002 and, as of 2012, has 114 automated stations at 107 locations in the lower 48 states, 12 stations in Alaska, and 2 stations in Hawai'i (Fig. 1). USCRN is operated by the National Oceanic and Atmospheric Administration and was designed to obtain long-term, accurate, and unbiased observations to quantify climate change on a national scale [9]. As such, the network follows strict requirements for site location, instrumentation, maintenance, documentation, and data storage and processing. With the aid of local experts, site locations were selected to obtain a minimum 50-year record without relocation and minimal changes in land use/land cover; hence, sites are primarily located on public lands (e.g., federal lands, university research stations, conservation areas) [9]. Sites are representative of the region and, to the extent possible, are situated in open areas with sufficient fetch (Fig. 2). Engineers calibrate all instruments on a regular schedule to the standards of the National Institute for Standards and Technology and both technicians and engineers maintain and document (e.g., via photographs) the site and its instrumentation during annual visits [9].

Instrumentation meets or exceeds standards set by the World Meteorological Organization and, as a key feature of the network, instrumentation for triple redundancy was installed for air temperature, precipitation, and, beginning in 2009, soil moisture and temperature (at 5, 10, 20, 50, and 100 cm). The redundancy ensures continuity of the climate record should an instrument fail and the enhanced quality assurance of all data, including better detection of instrument drift [9]. In addition to the redundant core measurements, stations also measure relative humidity, wind speed, skin (ground surface) temperature, and incoming solar radiation [10].

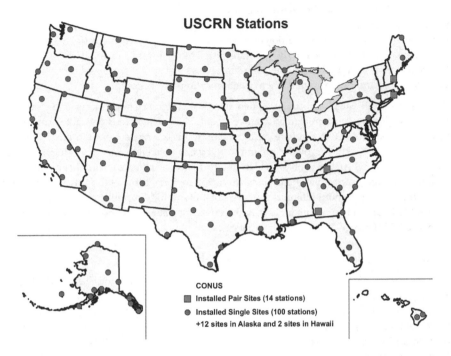

Figure 1. Map of the station distribution of U.S. Climate Reference Network at the end of 2012, including paired sites (blue squares) and individual sites (red dots) [9].

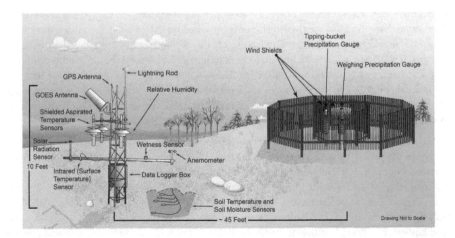

Figure 2. Schematic of the instrumentation at a typical U.S. Climate Reference Network station [9].

All information about the site, sensors, quality assurance procedures, data transmission and processing, and other pertinent information is documented as metadata and retained in perpetuity [9]. Five-minute-resolution observations of all variables except 10-cm and deeper soil observations (which are hourly) are transmitted every hour from all USCRN sites. Data that cannot be transmitted because of communications disruption are stored for up to 7 days at the local site and retransmitted when communications are restored.

Obviously, the climate monitoring provided by climate reference networks is highly desired; however, it also is a substantial financial investment. The investment is critical to providing a baseline climate record to compare to and calibrate observations from the other types of networks (e.g., [11, 12]), to provide "ground truth" data for satellite remote sensing of climate variables (e.g., [13, 14]), and to obtain a high-quality historical climate record that can measure long-term trends as well as variability and extremes in regional climates (e.g., [10, 15]). Because of multi-million and multi-billion dollar decisions made by the public and private sectors related to climate and, most importantly, its extremes, the value of a climate reference network such as USCRN greatly exceeds its financial costs.

3.2. Government-operated manual observation networks

As noted earlier, the historical climate record of the United States from the late 1800s to present results from manual observations from hundreds of Federal employees at what are commonly called "first-order stations" and thousands of trained volunteers in the U.S. Cooperative Observer Program [5], commonly called the COOP Network. In particular, a subset of the first-order and COOP stations — the U.S. Historical Climatology Network (HCN) — has been used to quantify national- and regional-scale temperature changes in the contiguous United States (Fig. 3) [16]. The 1,218 HCN stations (as of 2012) were selected based on their length of record (i.e., 80 years of monthly temperature and precipitation data), low percent of missing data, few or no station moves, and spatial coverage across the conterminous United States [17].

Because the stations were not chosen for their pristine locations or the perfection of their observational methods, several corrections were employed to adjust for systematic, non-climatic changes, resulting in the publication of the "United States Historical Climatology Network (HCN) Serial Temperature and Precipitation Data" record that is used worldwide by researchers, practitioners, and educators. These adjustments help to correct the time-of-observation bias [19, 20], station moves and instrument changes [21, 22], and undocumented discontinuities [23] such as incorrectly documenting the observation date, improperly resetting the maximum/minimum thermometer, or transcription errors [12].

Although it is desirable for all measurements to be consistent with the 10 climate monitoring principles proposed by Karl et al. (1995), the sheer number of manual observations from volunteers in the COOP Network precludes a pristine dataset. Yet, the COOP Network's data are invaluable as an historical record. From rural farmers to state climatologists to climate change researchers to policymakers, the COOP and HCN datasets are part of their toolkit for analysis and decision making. Unfortunately, as observers retire, move, or pass away, stations close and are rarely replaced with new ones. In addition, with the onset of state and private weather networks as well as shrinking budgets, local National Weather Service offices that

Cooperative Observer Program (COOP) Network

Figure 3. Distribution of COOP stations in the contiguous United States (black dots) and the U.S. Historical Climatology Network version 2 sites (red triangles) [18].

oversee their local COOP observing sites are rarely compelled to ensure that daily observations of temperature and precipitation from COOP sites receive high priority. The withering of the COOP Network is detrimental to our nation's historical climate record and, like many reductions in scientific observations, may not be recognized until it is too late to take action.

3.3. Automated mesoscale weather and climate monitoring networks

During the 1980s and 1990s, rapid advances in microcomputers and telecommunications led to the ability of states, private companies, educational institutions, nonprofit organizations, and other weather- and climate-sensitive groups to install and operate regional to national networks of weather or climate monitoring stations at a reasonable cost (e.g, $1,000 to $20,000 per station) [10]. Most of the early networks were to aid weather forecasting, with many stations established for K-12 education or for advertisement on the local news broadcasts (e.g., a company could invest in a station and have their name mentioned by on-air meteorologists or weatherpersons). More recently, thousands of small businesses, cities, water management agencies, agricultural producers, colleges, universities, and large corporations have invested in weather or climate monitoring. The rapid growth of these more affordable surface observing stations led, in many cases, to an emphasis on quantity over quality. Thousands of stations were installed on rooftops, under trees, near air conditioning units, and in other unsuitable locations.

There were numerous network operators, however, who took great care in design and implementation of their networks. Even before Karl et al.'s 10 principles of climate monitoring,

for example, scientists at the University of Oklahoma and Oklahoma State University established the Oklahoma Mesonet with adherence to standardization and principles to obtain high-
quality observations [24, 25]. Funded in late 1990 and commissioned in January 1994, the
Oklahoma Mesonet has 120 stations (in 2012, Fig. 4) that measure more than 20 environmental
variables, including air and soil temperatures (at multiple heights and depths), rainfall, relative
humidity, barometric pressure, solar radiation, wind speed (at multiple heights) and direction,
and soil moisture (at multiple depths) [15]. Stations transmit data every five minutes to a central
facility at the Oklahoma Climatological Survey, where observations immediately undergo a
suite of automated quality assurance tests [15, 26]. One or more meteorologists also manually
analyze the daily data as well as examine long-term trends or biases through a rigorous quality
assurance process [15]. Because the managing principle for the network from its inception was
"research-quality data in near real time," the Oklahoma Mesonet has served both the weather
and climate communities with great distinction for two decades, being called the "gold
standard of statewide mesoscale surface networks" by the National Research Council [10].

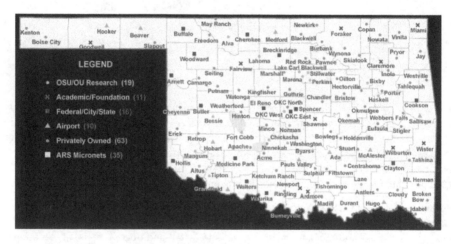

Figure 4. Location of Oklahoma Mesonet sites in 2012. The legend indicates land ownership of the sites and the two
yellow regions are non-Mesonet sites owned by the Agriculture Research Service that are operated and maintained by
the Oklahoma Mesonet. Image courtesy of the Oklahoma Climatological Survey.

Primarily developed and maintained by university researchers and staff, the number of
statewide or regional automated networks has been growing since the 1990s. Examples of these
multi-purpose networks include the following: West Texas Mesonet (70 stations in 2012, [27]),
New Jersey Weather and Climate Network (55 stations, http://climate.rutgers.edu/njwxnet/),
and Kentucky Mesonet (64 stations, [28, 29]). Within the private sector, networks such as Earth
Networks' Weather Network (http://www.earthnetworks.com/), WeatherFlow's Coastal
Mesonet (http://www.weatherflow.com/), and Ameren's Quantum Weather Mesonet (http://
www.ameren.com) have collaborated or partnered with university or government organizations to develop and operate quality commercial surface observing systems[1].

3.4. Use-inspired surface observing networks

Many regional surface observing systems were established initially to serve agricultural users and, in most cases, were expanded to serve a broader user community as additional funding or technology became available. Other examples of systems inspired by user needs include the following: off-shore platform measurement of weather and sea-surface conditions for the oil and gas industry, tall-tower observations of wind for the wind energy industry, and remote automated weather stations to observe potential wildfire conditions for fire management agencies.

Stations located in rural agricultural areas also are likely to be representative of a larger region and, hence, make excellent climate observing stations if measurement, metadata, and reporting standards are adhered to. Other use-inspired networks, however, have not been deemed acceptable for climate records, primarily because of station placement or the lack of metadata and standardization. For example, most state departments of transportation, particularly those that experience snow and ice conditions during the winter, installed road weather stations along major highways and interstates to monitor atmospheric and pavement conditions for road maintenance and safety. To satisfy operational needs, these network operators installed "environmental sensor stations" adjacent to pavement, many times in valleys, where road conditions would become hazardous sooner than flat, open areas. Hence, these stations only represented the microclimate of the area.

Until the 2000s, state road weather networks were rarely used by those outside the transportation community, and local and state network operators could not access each others' observations. To address challenges raised by the National Research Council in its report "Where the Weather Meets the Road: A Research Agenda for Improving Road Weather Services"   [30], transportation engineers, meteorologists, climatologists, systems engineers, and practitioners finally partnered in the Clarus Initiative to develop an integrated road weather observational network and data management system [31]. The initiative resulted in establishing guidelines for siting stations, documenting and maintaining metadata, developing quality assurance procedures, obtaining and sharing data in near real-time, and archiving data. The Federal Highway Administration and Intelligent Transportation System Joint Program Office of the U.S. Department of Transportation jointly administered the Clarus Initiative, prototyping the system in Fall 2006 for Alaska, Minnesota, and Utah and then incentivizing other states to join shortly thereafter. By 2012, 38 U.S. states and four Canadian provinces participated in the Clarus system. Because of the short period-of-record, Clarus data currently are not used for climate analyses; however, with continued adherence to standards and vigilance in operational oversight, the Clarus observations could become a particularly useful dataset to examine climate extremes, microclimates, and climates in mountainous terrain.

3.5. Citizen science networks

Taking advantage of the enthusiasm of weather hobbyists, school children, and those who want to learn about science, "citizen science" networks have grown over the years. The most

---

1 Reference to particular networks does not constitute an endorsement by the author or the University of Oklahoma.

expansive of these grassroots efforts, with over 15,000 volunteers in 2010, is the Community Collaborative Rain, Hail and Snow Network (CoCoRaHS) [32, 33]. Founded in 1998, CoCoR-aHS provides a platform for communities around the United States to participate in "hands-on" science in their own backyards through the daily observation of precipitation. Volunteers of all ages measure rain, hail, and snow using simple, low-cost rain gauges and hail pads. Observers input their measurements through the CoCoRaHS web site (www.cocorahs.org), where data become immediately available. These data are used daily by many federal, state, and community organizations. Through consistent observation and data sharing, participants actively learn about their climate, their water resources, and the impacts precipitation has on their lives and their communities. In some areas of the country, there is a greater concentration of observers (e.g., south-central Texas, Colorado's Front Range, metropolitan Chicago and Phoenix) where the density is approaching one station per 2.5 to 3 square kilometers, offering detailed insight into precipitation variability and extremes.

4. Benefits and limitations of surface observing of the climate

As federal budgets for environmental observing systems and numerical modeling become tighter, there has been a tendency for modelers and remote sensing experts to state that their technologies are sufficient to monitor changes in the environment. In the case of satellite remote sensing of weather and climate variables, which is necessary for global observations particu-larly across mountainous terrain, oceans, and countries with few or classified surface obser-vations, any under-appreciation for surface observing systems could become detrimental to satellite programs themselves as well as our nation's historical climate record. "Ground truth" of satellite products — derived from empirical relationships between the desired product (e.g., air temperature) and the measured radiances — requires a long-term record of high-quality surface observations. This dataset is needed not only to develop satellite algorithms but to ensure that long-term climatic trends, synthesized from multiple satellite instruments that have changed significantly from the 1960s to present, are consistent with those detected over land during the same time period.

Both active and passive remote sensing have distinct advantages in their spatial coverage over point-based surface observations, especially in regions such as the American West and Alaska, where surface observing stations are relatively sparse compared to the more populated or flatter regions of the United States [34]. For example, dual-polarimetric weather radars can identify the type and amount of hydrometeors within the troposphere with a resolution high enough to capture intense precipitation shafts that may result in flash flooding (e.g., [35, 36] or distinguish between winter precipitation types (e.g., rain, snow, and sleet; [37]). Small-scale precipitation maxima can move between surface observing sites and, thus, not be measured, and most surface stations are not equipped with a present weather sensor to determine the precipitation type. In addition, satellite imagers can take advantage of radiation bands (e.g., infrared, thermal, microwave) to detect climate-related changes in the environment (e.g., phenology) not measured by surface observing stations (e.g., [34, 38-41]).

It is clearly important to measure variables throughout the full vertical profile of the atmosphere, especially the boundary layer, so remotely sensed observations are critical; however, most environmental, economic, and societal decisions are made based on land surface climatology, from a meter or so below the ground surface to several meters above it. Within this region, businesses and residences exist, much of the biosphere evolves, people and goods are transported, wildlife migrates, and water moves and undergoes phase changes. Policymakers also affect public policy (e.g., flood insurance, agricultural incentives, protection of natural resources, emissions limitations) as a result of conditions and changes primarily near the land surface.

The most accurate measurements of the Earth's land surface are in-situ observations because satellite and radar remote sensing have significant limitations. In-situ, not satellite or radar, sensors can measure the atmospheric pressure gradients that drive weather systems. In-situ sensors can measure variables regardless of whether there is cloud cover or not, at precise heights or depths, and with high temporal resolution — all which are currently unachievable with satellite remote sensing (e.g., [34, 42]). In-situ networks provide the validation data needed for remote sensing and are inexpensive compared to satellite instrumentation, sensor deployment, and sensor maintenance and replacement costs (e.g., [10, 43]). In most cases, a broken or poorly functioning in-situ sensor can be replaced by a well-calibrated twin within days, leading to a more continuous data record (e.g., [15]). Because of their high temporal resolution, in-situ networks measure climate extremes and the full range of variability (e.g., [44, 45]) that is needed for local to international decision making. For the historical climate record and analysis of important trends in a changing climate, in-situ sensors observe air and soil temperature, precipitation, air and soil moisture, and winds directly whereas satellite imaging, for example, averages radiances over a large footprint, suffers from atmospheric attenuation, and has poor vertical resolution; thus, satellites cannot adequately measure low-level atmospheric moisture, surface winds, or precipitation (e.g., [46-48]).

Mesoscale surface observing systems are not without their limitations however. Several problems are characteristic of most surface observing sites: change in land use/land cover over time, necessary relocation or closure of sites, increasing maintenance with sensor age, inadequate spatial coverage (especially for precipitation), limited length of record, differing times of observation, and poorly documented and maintained metadata [9]. With adequate standards and attention to data quality assurance, these problems can be significantly reduced in any given network, allowing for research-quality time series at each station.

One of the greatest benefits of a high-resolution automated mesoscale climate network is the ability to observe rapidly evolving phenomena that are indicative of a region's climate. For example, Figure 5 displays a significant temperature change (~18°C in 30 min) that is indicative of the extremes of Alaska's climate. In this case, a warm air mass from the Gulf of Alaska penetrated inland in the Tetlin National Wildlife Refuge in southeast Alaska (near the U.S.–Canada border), relieving a lengthy cold stretch. To better engineer structures (e.g., above-ground pipelines) and prepare for health consequences of this type of harsh environment, high-resolution observations are needed.

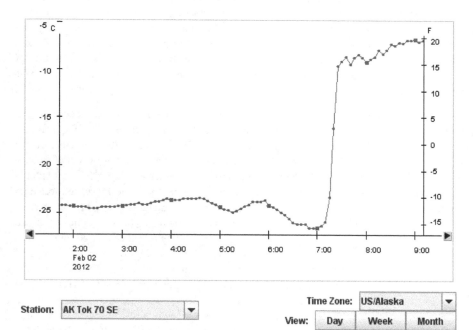

-●- AK Tok 70 SE - Calculated Temperature

Station: AK Tok 70 SE

Time Zone: US/Alaska

View: Day Week Month

Source: National Climatic Data Center/NESDIS/NOAA

Figure 5. Shortly after 7 AM local time on 2 February 2012, the air temperature at 1.5 meters above ground at the "Tok 70 SE" station (Alaska) increased 13.7°C in 10 min, and almost 18°C in 30 min. The observations documented a change in air masses in southeast Alaska that is part of the region's climate. Data courtesy of the U.S. Climate Reference Network (M. Palecki, personal communication).

A benefit of the high spatial resolution of a mesoscale monitoring network is the ability to document phenomena that typically occur between widely spaced synoptic stations. For example, decaying nocturnal thunderstorms can produce hot, dry, and gusty surface winds called "heatbursts," typically generated in an environment characterized by a dry boundary layer and surface temperature inversion [49]. Prior to the analysis of an 18-year record of Oklahoma Mesonet data, less than two dozen heatbursts had been documented worldwide in the scientific literature. With the high-resolution Mesonet observations however, 207 heatburst events[2] of various magnitudes, areal coverage, and duration were identified between 1994 and 2009 across Oklahoma [50]. The geographical distribution of heatbursts across Oklahoma (Fig. 6) documents the characteristically drier boundary layer in the western (versus the eastern)

---

2 A heatburst event was defined as "heatburst or short series of heatbursts that (1) affected one or more Mesonet stations during a single heatburst day and (2) demonstrated temporal and spatial continuity."

part of the state. A longer period of record of heatbursts should be able to document whether or not the regional boundary-layer moisture gradient shifts position with climate change.

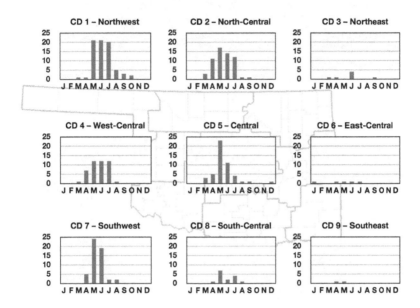

Figure 6. Geographical distribution of heatburst detections by climate division (CD) across Oklahoma and by month (J, F, M,... for January, February, March,...) from 1 January 1994 to 15 August 2009 [50]. If a heatburst event were measured at 10 Oklahoma Mesonet sites, it was counted as 10 detections.

The traditional separation of "weather observations" and "climate observations" is being blurred with the near real-time availability of high-quality, high-resolution data from automated mesoscale observing networks. These observations are used both by weather forecasters to protect lives and property during extreme events and by climatologists to analyze and document these extremes, their trends over time, and their impacts on physical, natural, and human systems. For example, the Kentucky Mesonet measured 5-min rainfall during a heavy precipitation event on 1-2 May 2010 in the Mid-South U.S. (Fig. 7). The larger event resulted in the historic flooding of downtown Nashville, Tennessee. The Mesonet station received 8.38 mm of rain during a 5-min period, 50.8 mm during an hour, and 258 mm over two days — the latter broke the state's two-day precipitation record of 211 mm (from 6-7 December 1924) [51]. Similarly, daily rainfall observations from multiple climate monitoring networks were used to request disaster assistance from the Federal Emergency Management Agency for damaging storms caused by the rare re-intensification of a tropical storm (Erin) over western Oklahoma on 19 August 2007 (Fig. 8) [52]. As another example, the New Jersey Weather and Climate Network provided high-resolution data (Fig. 9) for government officials and the public before, during, and after Hurricane Sandy (2012), which devastated the Northeast U.S., especially coastal areas of New Jersey.

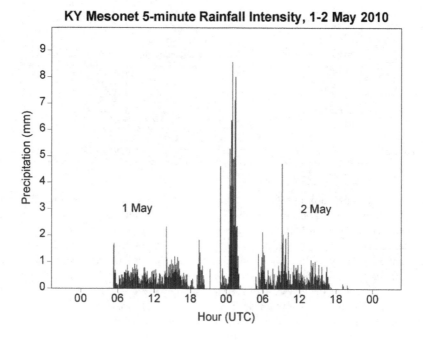

Figure 7. Five-minute rainfall (in millimeters) at Bowling Green, Kentucky, during the 1–2 May 2010 flooding event [51].

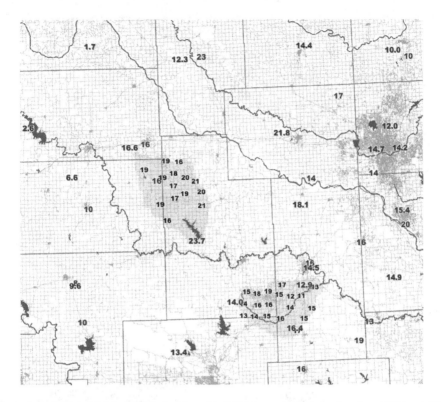

Figure 8. Storm-total rainfall (cm) for Tropical Storm Erin (2007) across west-central and southwest Oklahoma [52]. Values were measured by the Oklahoma Mesonet (dark red), the Fort Cobb and Little Washita Micronets of the Agriculture Research Service (rainfall values in black, micronet domain shaded in light green; Fort Cobb is northwest of the Little Washita), and the U.S. Cooperative Observer Network (dark green).

Mesoscale observations have been combined with remotely sensed data to produce important products for decision makers that could not be developed with either type of observation alone. For example, to depict the potential for wildfire, the Oklahoma Fire Danger Model [53] includes a set of maps that combines temperature, relative humidity, and wind observations from the Oklahoma Mesonet with Normalized Difference Vegetation Index product (1-km resolution) of polar orbiting satellites as well as dead fuel models. Map products include a burning index (Fig. 10) that represents the estimated intensity of any fire that occurs, an ignition component that estimates how easily a firebrand can produce a fire, a spread component that represents the forward speed of any headfire, and an energy release component that estimates the heat released per unit area at the head of a fire. Firefighters use these products to staff their teams in advance of days with high fire danger and protect both their teams and the public during wildfire events. This example demonstrates the benefits of developing robust remote sensing and in-situ observational networks in tandem so as to develop decision products that save lives, property, and money.

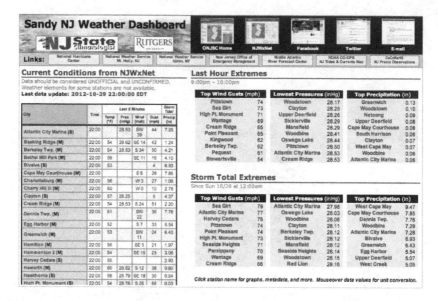

Figure 9. An online "dashboard" established by the New Jersey Weather and Climate Network for decision makers and the public during Hurricane Sandy (2012). This special web page provided 5-min updates from 52 observing stations in and near New Jersey. It was used by 30,000 unique visitors during the storm and was posted continuously at the New Jersey Emergency Operations Center (D. Robinson, personal communication).

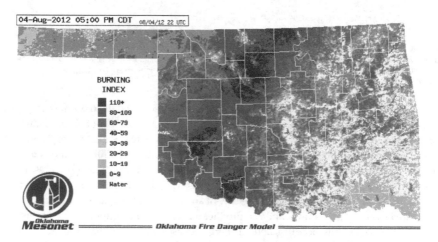

Figure 10. Example of the burning index (BI) product from the Oklahoma Fire Danger Model [53]. The index directly relates to the intensity of the fire (hence, the difficulty of containment) and is scaled such that BI/10 is equal to the flame length (in feet) at the head of the fire. The Oklahoma Fire Danger Model results from collaborations of Oklahoma State University, the University of Oklahoma, and the Fire Sciences Laboratory of the USDA Forest Service in Missoula, MT.

5. What might the future hold for surface observing of the climate?

In 2009, the National Research Council (NRC) issued a report documenting the status of weather and climate observing systems in the United States and recommended actions to evolve the existing, disparate networks into an integrated, coordinated "network of networks" [10]. The NRC report acknowledged the pivotal role of the federal government, especially the National Oceanic and Atmospheric Administration, in weather and climate monitoring and information services, but also recognized the rapidly growing roles of state and local governments, universities, and the private sector in obtaining mesoscale observations. The broader weather and climate enterprise was investing in surface observing; however, there was little coordination, standardization, and leadership provided to ensure that a system was created to be greater than the sum of its individual parts.

In response to the 2009 NRC report, the American Meteorological Society, whose membership comprises governmental, private sector, and academic interests in the weather and climate enterprise, established an Ad Hoc Committee on a Nationwide Network of Networks. Through six subcommittees with representatives from across the enterprise, the committee provided a vision for a path forward from the existing disarray in mesoscale observation systems [54]. A steering group comprised of experts from the public, private, and academic sectors should be created to adopt technical standards for participating networks, incentivize networks to participate, coordinate activities of the nationwide network of networks (e.g., data sharing policies and costs, product development and dissemination, participant workshops and educational activities), and develop the market for mesoscale observations [54].

These and other reports have highlighted the fact that, especially in times of austerity and other fiscal pressures, the federal government cannot alone be responsible for monitoring the nation's climate. With considerable funding and interest in surface observing from non-federal entities with significant experience and expertise, it would be wise for the federal government to recognize the value of providing leadership in coordinating the growing array of networks. Hand in hand with this coordination effort, the federal government should work to integrate its own suite of surface observing systems and prioritize highly both a climate reference network (to which other observing systems can be compared for quality) and the historical cooperative observer network (for long-term continuity). It is anticipated that success in creating a nationwide network of networks would generate commercial revenue sufficient to help reduce the costs to the government for their observation systems. Imagine a Google, Inc. model to obtaining and sharing weather and climate data; the income generated could revolutionize the financial structure of surface observations.

A dramatic transformation is indeed needed. A broader vision for "research quality data in near real-time" must evolve and permeate the enterprise to minimize the competition between weather monitoring for public safety and climate monitoring for climate change assessment. In particular, although mistakes have been made in the past by the private sector, the climate enterprise should embrace an appropriate role for the private sector as leaders in innovation, cost efficiency, and marketing. Perhaps a certification program, overseen by independent experts, can raise the value of excellence in environmental monitoring to ensure a growing record of climate observations for decades to come.

6. Conclusion

The United States has a rich history of monitoring the climate using surface observations, from thousands of dedicated volunteer observers to governmental, academic, and commercial networks using automated stations and reliable telecommunications. In the past several decades, these observations have been augmented by remote sensing technologies that provide necessary spatial coverage and the ability to monitor non-atmospheric variables related to climate variability and change. Hand in hand, in-situ and remotely sensed measurements provide data for research, education, and daily decision making that many times are undervalued by funding agencies, especially as budgets tighten.

As we struggle with the consequences of climate change, surface observing systems become more critical to monitor how climate variability, especially extremes in temperature and precipitation, evolves across different geographical regions. These observations have become critical components not only in establishing our climate record, but in ensuring that global and regional climate models represent the key climate drivers across the world as well as providing routine, hourly or sub-hourly observations for daily decisions by natural resource managers, public safety officials, transportation managers, agricultural producers, event planners, weather forecasters, and others. A robust and healthy surface observing system — from climate reference networks to regional mesonets to volunteer observers — must be maintained for the security of national, regional, and local economies and the protection of our natural resources.

Acknowledgements

I sincerely appreciate the help of Bruce Baker (NOAA's National Climatic Data Center), Howard Diamond (NOAA's National Climatic Data Center), Rezaul Mahmood (Western Kentucky University), William McPherson (University of Oklahoma), Henry Reges (Colorado State University), David Robinson (Rutgers University), and Michael Palecki (NOAA's National Climatic Data Center) for their quick response to my requests for information.

Author details

Renee A. McPherson

Address all correspondence to: renee@ou.edu

Department of Geography and Environmental Sustainability, University of Oklahoma, Norman, Oklahoma, USA

References

[1] Slonosky, V.C., 2002: Wet winters, dry summers? Three centuries of precipitation data from Paris. Geophys. Res. Lett., 29: 1895.

[2] Manley, G., 1953: The mean temperature of central England, 1698–1952. Quarterly Journal of the Royal Meteorological Society, 79: 242-261.

[3] Fiebrich, C.A., 2009: History of surface weather observations in the United States. Earth-Science Reviews, 93: 77-84.

[4] Davies, A. and O.M. Ashford, 1990: Forty years of progress and achievement: a historical review of WMO: Secretariat of the World Meteorological Organization pp.

[5] National Weather Service, 2012: Cooperative Program Management and Operations. NWS Instruction 10-1307. Department of Commerce, Editor. National Oceanic & Atmospheric Administration: Silver Spring, MD, 54 pp.

[6] Eden, P., 2009: Traditional weather observing in the UK: An historical overview. Weather, 64: 239-245.

[7] National Research Council, 2012: The National Weather Service Modernization and Associated Restructuring: A Retrospective Assessment. The National Academies Press: Washington, D.C., 140 pp.

[8] Karl, T.R., V.E. Derr, D.R. Easterling, C.K. Folland, D.J. Hoffman, S. Levitus, N.Nicholls, D.E. Parker, and G.W. Withee, 1995: Critical issues for long-term climate monitoring. Climatic Change, 31: 185-221.

[9] Diamond, H.J., T.R. Karl, M.A. Palecki, C.B. Baker, J.E. Bell, R.D. Leeper, D.R. Easterling, J.H. Lawrimore, T.P. Meyers, M.R. Helfert, G. Goodge, and P.W. Thorne, 2012: U.S. Climate Reference Network after one decade of operations: Status and assessment. Bulletin of the American Meteorological Society, In press.

[10] National Research Council, 2009: Observing Weather and Climate from the Ground Up: A Nationwide Network of Networks, ed. Committee on Developing Mesoscale Meteorological Observational Capabilities to Meet Multiple National Needs. Washington, D.C.: The National Academies Press, 234 pp.

[11] Menne, M.J., C.N. Williams, Jr., and M.A. Palecki, 2010: On the reliability of the U.S. surface temperature record. J. Geophys. Res., 115: D11108.

[12] Fiebrich, C.A. and K.C. Crawford, 2009: Automation: A step toward Improving the quality of daily temperature data produced by climate observing networks. Journal of Atmospheric and Oceanic Technology, 26: 1246-1260.

[13] Guillevic, P.C., J.L. Privette, B. Coudert, M.A. Palecki, J. Demarty, C. Ottlé, and J.A. Augustine, 2012: Land Surface Temperature product validation using NOAA's sur-

face climate observation networks—Scaling methodology for the Visible Infrared Imager Radiometer Suite (VIIRS). Remote Sensing of Environment, 124: 282-298.

[14] Collow, T.W., A. Robock, J.B. Basara, and B.G. Illston, 2012: Evaluation of SMOS retrievals of soil moisture over the central United States with currently available in situ observations. J. Geophys. Res., 117: D09113.

[15] McPherson, R.A., C.A. Fiebrich, K.C. Crawford, R.L. Elliott, J.R. Kilby, D.L. Grimsley, J.E. Martinez, J.B. Basara, B.G. Illston, D.A. Morris, K.A. Kloesel, S.J. Stadler, A.D. Melvin, A.J. Sutherland, H. Shrivastava, J.D. Carlson, J.M. Wolfinbarger, J.P. Bostic, and D.B. Demko, 2007: Statewide monitoring of the mesoscale environment: A technical update on the Oklahoma Mesonet. Journal of Atmospheric and Oceanic Technology, 24: 301-321.

[16] Menne, M.J., C. N. Williams Jr., and R.S. Vose, 2009: The U.S. Historical Climatology Network Monthly Temperature Data, Version 2. Bulletin of the American Meteorological Society, 90: 993-1007.

[17] Karl, T.R., C. N. Williams Jr., and F.T. Quinlan, 1990: United States Historical Climatology Network (HCN) Serial Temperature and Precipitation Data. ORNL/CDIAC-30, NDP-019/R1. Carbon Dioxide Information Analysis Center, Oak Ridge National Laboratory, U.S. Department of Energy: Oak Ridge, TN, 392 pp.

[18] C. N. Williams Jr., M.J. Menne, and P.W. Thorne, 2012: Benchmarking the performance of pairwise homogenization of surface temperatures in the United States. J. Geophys. Res., 117: D05116.

[19] Baker, D.G., 1975: Effect of Observation Time on Mean Temperature Estimation. Journal of Applied Meteorology, 14: 471-476.

[20] Karl, T.R., C.N. Williams, P.J. Young, and W.M. Wendland, 1986: A Model to Estimate the Time of Observation Bias Associated with Monthly Mean Maximum, Minimum and Mean Temperatures for the United States. Journal of Climate and Applied Meteorology, 25: 145-160.

[21] Karl, T.R. and C.N. Williams, 1987: An Approach to Adjusting Climatological Time Series for Discontinuous Inhomogeneities. Journal of Climate and Applied Meteorology, 26: 1744-1763.

[22] Quayle, R.G., D.R. Easterling, T.R. Karl, and P.Y. Hughes, 1991: Effects of Recent Thermometer Changes in the Cooperative Station Network. Bulletin of the American Meteorological Society, 72: 1718-1723.

[23] Menne, M.J. and C.N. Williams, 2005: Detection of Undocumented Changepoints Using Multiple Test Statistics and Composite Reference Series. Journal of Climate, 18: 4271-4286.

[24] Brock, F.V., K.C. Crawford, R.L. Elliott, G.W. Cuperus, S.J. Stadler, H.L. Johnson, and M.D. Eilts, 1995: The Oklahoma Mesonet: A technical overview. Journal of Atmospheric and Oceanic Technology, 12: 5-19.

[25] Shafer, M.A., C.A. Fiebrich, D.S. Arndt, S.E. Fredrickson, and T.W. Hughes, 2000: Quality assurance procedures in the Oklahoma Mesonetwork. Journal of Atmospheric and Oceanic Technology, 17: 474-494.

[26] Fiebrich, C.A., C.R. Morgan, A.G. McCombs, P.K. Hall, and R.A. McPherson, 2010: Quality assurance procedures for mesoscale meteorological data. Journal of Atmospheric and Oceanic Technology, 27: 1565-1582.

[27] Schroeder, J.L., W.S. Burgett, K.B. Haynie, I. Sonmez, G.D. Skwira, A.L. Doggett, and J.W. Lipe, 2005: The West Texas Mesonet: A Technical Overview. Journal of Atmospheric and Oceanic Technology, 22: 211-222.

[28] Grogan, D.M., 2010: Information technology implementation decisions to support the Kentucky Mesonet. Publications in Climatology: 1-106.

[29] Grogan, D.M., S. Foster, and M. R., 2010: The Kentucky Mesonet: Perspectives on data access, distribution, and use for a mesoscale surface network. Proceedings of the 26th Conference on Interactive Information and Processing Systems (IIPS) for Meteorology, Oceanography, and Hydrology: Boston, MA.

[30] National Research Council, 2004: Where the Weather Meets the Road: A Research Agenda for Improving Road Weather Services, ed. Committee on Weather Research for Surface Transportation: The Roadway Environment. Washington, D.C.: The National Academies Press, 188 pp.

[31] Pisano, P.A., P.J. Kennedy, and A.D. Stern (2008) A New Paradigm in Observing the Near Surface and Pavement: Clarus and Vehicle Infrastructure Integration. Transportation Research E-Circular E-C126, 3-15.

[32] Cifelli, R., N. Doesken, P. Kennedy, L.D. Carey, S.A. Rutledge, C. Gimmestad, and T. Depue, 2005: The Community Collaborative Rain, Hail, and Snow Network: Informal Education for Scientists and Citizens. Bulletin of the American Meteorological Society, 86: 1069-1077.

[33] Doesken, N. and H. Reges, 2010: The Value of the Citizen Weather Observer. Weatherwise, November-December: 30-37.

[34] Kidd, C., V. Levizzani, and P. Bauer, 2009: A review of satellite meteorology and climatology at the start of the twenty-first century. Progress in Physical Geography, 33: 474-489.

[35] Sachidananda, M. and D.S. Zrnić, 1987: Rain Rate Estimates from Differential Polarization Measurements. Journal of Atmospheric and Oceanic Technology, 4: 588-598.

[36] Chandrasekar, V., H. Chen, and M. Maki, 2012: Urban flash flood applications of high-resolution rainfall estimation by X-band dual-polarization radar network. Re-

mote Sensing of the Atmosphere, Clouds, and Precipitation IV: Kyoto, Japan, October 29, 2012.

[37] Schuur, T.J., H.-S. Park, A.V. Ryzhkov, and H.D. Reeves, 2012: Classification of Precipitation Types during Transitional Winter Weather Using the RUC Model and Polarimetric Radar Retrievals. Journal of Applied Meteorology and Climatology, 51: 763-779.

[38] Vrieling, A., K. Beurs, and M. Brown, 2011: Variability of African farming systems from phenological analysis of NDVI time series. Climatic Change, 109: 455-477.

[39] White, M.A., K.M. De Beurs, K. Didan, D.W. Inouye, A.D. Richardson, O.P. Jensen, J. O'Keefe, G. Zhang, R.R. Nemani, W.J.D. Van Leeuwen, J.F. Brown, A. De Wit, M. Schaepman, X. Lin, M. Dettinger, A.S. Bailey, J. Kimball, M.D. Schwartz, D.D. Baldocchi, J.T. Lee, and W.K. Lauenroth, 2009: Intercomparison, interpretation, and assessment of spring phenology in North America estimated from remote sensing for 1982–2006. Global Change Biology, 15: 2335-2359.

[40] Ge, J., 2010: MODIS observed impacts of intensive agriculture on surface temperature in the southern Great Plains. International Journal of Climatology, 30: 1994-2003.

[41] Liang, L. and M. Schwartz, 2009: Landscape phenology: an integrative approach to seasonal vegetation dynamics. Landscape Ecology, 24: 465-472.

[42] Folland, C.K., T.R. Karl, and M. Jim Salinger, 2002: Observed climate variability and change. Weather, 57: 269-278.

[43] Kramer, H.J., 2002: Observation of the Earth and Its Environment: Survey of Missions and Sensors: Springer Verlag pp.

[44] Alexander, L.V., X. Zhang, T.C. Peterson, J. Caesar, B. Gleason, A.M.G. Klein Tank, M. Haylock, D. Collins, B. Trewin, F. Rahimzadeh, A. Tagipour, K. Rupa Kumar, J. Revadekar, G. Griffiths, L. Vincent, D.B. Stephenson, J. Burn, E. Aguilar, M. Brunet, M. Taylor, M. New, P. Zhai, M. Rusticucci, and J.L. Vazquez-Aguirre, 2006: Global observed changes in daily climate extremes of temperature and precipitation. Journal of Geophysical Research: Atmospheres, 111: n/a-n/a.

[45] Easterling, D.R., G.A. Meehl, C. Parmesan, S.A. Changnon, T.R. Karl, and L.O. Mearns, 2000: Climate Extremes: Observations, Modeling, and Impacts. Science, 289: 2068-2074.

[46] Sharma, N. and M.M. Ali, 2013: A Neural Network Approach to Improve the Vertical Resolution of Atmospheric Temperature Profiles From Geostationary Satellites. Geoscience and Remote Sensing Letters, IEEE, 10: 34-37.

[47] Parekh, A., R. Sharma, and A. Sarkar, 2007: A Comparative Assessment of Surface Wind Speed and Sea Surface Temperature over the Indian Ocean by TMI, MSMR, and ERA-40. Journal of Atmospheric and Oceanic Technology, 24: 1131-1142.

[48] Sapiano, M.R.P. and P.A. Arkin, 2009: An Intercomparison and Validation of High-Resolution Satellite Precipitation Estimates with 3-Hourly Gauge Data. Journal of Hydrometeorology, 10: 149-166.

[49] Johnson, B.C., 1983: The Heat Burst of 29 May 1976. Monthly Weather Review, 111: 1776-1792.

[50] McPherson, R.A., J.D. Lane, K.C. Crawford, and W.G. McPherson, 2011: A climatological analysis of heatbursts in Oklahoma (1994-2009). International Journal of Climatology, 31: 531-544.

[51] Durkee, J.D., L. Campbell, K. Berry, D. Jordan, G. Goodrich, R. Mahmood, and S. Foster, 2011: A Synoptic Perspective of the Record 1-2 May 2010 Mid-South Heavy Precipitation Event. Bulletin of the American Meteorological Society, 93: 611-620.

[52] Arndt, D.S., J.B. Basara, R.A. McPherson, B.G. Illston, G.D. McManus, and D.B. Demko, 2009: Observations of the overland reintensification of Tropical Storm Erin (2007). Bulletin of the American Meteorological Society, 90: 1079-1093.

[53] Carlson, J.D., R.E. Burgan, D.M. Engle, and J.R. Greenfield, 2002: The Oklahoma Fire Danger Model: An operational tool for mesoscale fire danger rating in Oklahoma. International Journal of Wildland Fire, 11: 183-191.

[54] Frederick, G.L., P. Campbell, J. Lasley, R.A. McPherson, R. Pasken, B. Philips, and J. Stalker, 2011: Report of the AMS ad hoc Committee on a Nationwide Network of Networks. Final Draft. American Meteorological Society: Washington, DC, 94 pp.

Chapter 8

The Effect of Agricultural Growing Season Change on Market Prices in Africa

K. M. de Beurs and M. E. Brown

Additional information is available at the end of the chapter

1. Introduction

Local agricultural production is a key element of food security in many agricultural countries in Africa. Climate change and variability is likely to adversely affect these countries, particularly as they affect the ability of smallholder farmers to raise enough food to feed themselves. Seasonality influences farmers' decisions about when to sow and harvest, and ultimately the success or failure of their crops.

At a 2009 conference in the United Kingdom hosted by the Institute of Development Studies, Jennings and Magrath (2009) described farmer reports from East Asia, South Asia, Southern Africa, East Africa and Latin America. Farmers indicate significant changes in the timing of rainy seasons and the pattern of rains within seasons, including:

- More erratic rainfall, coming at unexpected times in and out of season;

- Extreme storms and unusually intense rainfall are punctuated by longer dry spells within the rainy season;

- Increasing uncertainty as to the start of rainy seasons in many areas;

- Short or transitional second rainy seasons are becoming stronger than normal or are disappearing altogether.

These farmer perceptions of change are striking in that they are geographically widespread and are remarkably consistent across diverse regions (Jennings and Magrath, 2009). The impact of these changes on farmers with small plots and few resources is large. Farming is becoming riskier because of heat stress, lack of water, pests and diseases that interact with ongoing pressures on natural resources. Lack of predictability in the start and length of the growing season affects the ability of farmers to invest in appropriate fertilizer levels or improved, high yielding varieties. These changes occur at the same time as the demand for food is rising and is projected to continue to rise for the next fifty years (IAASTD, 2008).

Long-term data records derived from satellite remote sensing can be used to verify these reports, providing necessary analysis and documentation required to plan effective adaptation strategies. Remote sensing data can also provide some understanding of the spatial extent of these changes and whether they are likely to continue.

Given the agricultural nature of most economies on the African continent, agricultural production continues to be a critical determinant of both food security and economic growth (Funk and Brown, 2009). Crop phenological parameters, such as the start and end of the growing season, the total length of the growing season, and the rate of greening and senescence are important for planning crop management, crop diversification, and intensification.

The World Food Summit of 1996 defined food security as: "when all people at all times have access to sufficient, safe, nutritious food to maintain a healthy and active life". Food security roughly depends on three factors: 1) availability of food; 2) access to food and 3) appropriate use of food, as well as adequate water and sanitation. The first factor is dependent on growing conditions and weather and climate. In a previous paper we have investigated this factor by evaluating the effect of large scale climate oscillation on land surface phenology (Brown et al., 2010). We found that all areas in Africa are significantly affected by at least one type of large scale climate oscillations and concluded that these somewhat predictable oscillations could perhaps be used to forecast agricultural production. In addition, we have evaluated changes in agricultural land surface phenology over time (Brown et al., 2012). We found that land surface phenology models, which link large-scale vegetation indices with accumulated humidity, could successfully predict agricultural productivity in several countries around the world.

In this chapter we are interested in the effect of variability in peak timing of the growing season, or phenology, on the second factor of food security, food access. In this chapter we want to determine if there is a link between market prices and land surface phenology and to determine which markets are vulnerable to land surface phenology changes and variability and which market prices are not correlated.

2. Background

2.1. Vegetation seasonality and change

Early research on the impact of global climate change on the growing season in northern latitudes was based on satellite remote sensing observations of vegetation (Myneni et al., 1997, Nemani et al., 2003, Slayback et al., 2003). These direct observations of change in the onset of spring led to the development of phenological models using remote sensing information. Phenology is the study of the timing of recurring biological cycles and their connection to climate (Lieth, 1974). Phenology has the promise of capturing quantitatively the changes reported by farmers and providing evidence for its link to climate change. Land surface phenology is the analysis of changes in the vegetated land surface as observed by satellite images (de Beurs and Henebry, 2004). Land surface phenology distinguishes itself from species centric phenology in that it focuses on the analysis of the land surface in mixed pixels. White et al. (2009) described the complexity of comparing ground observations of the start of season with satellite-derived estimates due to the difficulty in understanding the myriad definitions of season metrics.

Land surface phenology models rely on remote sensing information of vegetation, such as the dataset derived from the Advanced Very High Resolution Radiometer (AVHRR) (Tucker et al., 2005) and the newer MODIS sensors on Aqua and Terra. Vegetation and rainfall data can assess variables such as the start of season, growing season length and overall growing season productivity (Brown and De Beurs, 2008, Brown, 2008). These metrics are common inputs to crop models that estimate the impact of weather on yield (Verdin and Klaver, 2002). Land surface phenology metrics have a strong relationship with regional food production, particularly those with sufficiently long records to capture local variability (Funk and Budde, 2009, Vrieling et al., 2008).

2.2. Price seasonality

The integration of a market into the broader economy has been the objective of many development programs (Barrett, 2008), since the increased integration of food markets in developing countries is considered to be of vital importance for agricultural transformation and economic growth (Fafchamps, 1992). Market integration is also an important aspect of food security, since many sub-Saharan countries face food shortages as a result of crop failures caused by drought or other climatic hazards (Zant, 2013). Integrated markets offer the potential to reduce the impact of weather shocks by quickly moving food from surplus to deficit areas. Conversely, poorly integrated markets, such as those where inadequate trade infrastructure hinders market function, may result in food shortages (Zant, 2013). Poorly integrated markets are often isolated because of low participation in the market by farmers, resulting in 'thin' markets that have too little supply during times of high demand (before the harvest) and too much supply during times with low demand (after the harvest). Many households in developing countries seek to be as self-sufficient as possible in capital, labor and food to reduce exposure to variability in prices and extremely high transaction costs (Lutz et al., 1995), which are both a cause and a consequence of thinly traded, volatile markets. Thinly traded markets keep the difference between producer and consumer prices high, further reinforcing household incentives to minimize their reliance on markets (Tschirley and Weber, 1994, Kelly et al., 1996).

Seasonality in food prices, as measured by the ratio of post harvest to harvest prices, is high in markets that are poorly integrated and isolated. Seasonal price changes may reflect changes in production, particularly in good years when infrastructure and trade constraints reduce the ability of traders to move excess grain out of an area. Seasonal price spreads can be explained by storage losses, large postharvest grain sales, and lack of trader participation in isolated markets during average and good years (Alderman and Shively, 1996). Thus price seasonality is negatively related to production anomalies, where higher (lower) production will create lower (higher) prices during the post harvest season because of the inability or unwillingness of households and traders to store grain.

Another source of seasonality in food prices in thin markets is the seasonality of transaction costs, as well as transportation costs. Little is known about the variability of transportation costs in each of the markets of this study, but rainfall and poor roads, increased demand for movement of goods and people during the rainy season, and the increased difficulty of distributing fuel and other necessities for transportation make it likely that transportation costs will be higher during the growing seasons (Alderman and Shively, 1996). Transaction costs are the costs incurred in making an economic exchange: determining the price and the demand for a good in a market, the cost of bargaining for a fair price, and the cost of policing and enforcement in the market (Asante et al., 1989, Fafchamps, 2004). All of these costs are also likely to be seasonal. These sources of non-food production variability in the seasonality of food prices can also be estimated with remote sensing data.

3. Data

3.1. MODIS data

MODIS Nadir BRDF-Adjusted Reflectance data (Schaaf et al., 2002) at 0.05° spatial resolution with temporal resolution of 16 days (MCD43C4) and temporal coverage from 2001 through 2011 were employed to derive the Normalized Difference Vegetation Index as:

$$NDVI = (NIR - RED) / (NIR + RED) \tag{1}$$

Where, NIR is the near infrared reflectance of MODIS band 2 (841-876 nm) and RED is MODIS band 1 (620-670 nm). NDVI is an often used vegetation index (Brown et al., 2006, Karnieli et al., 2010) which exploits the significant difference between NIR and Red reflectance for living vegetation. Healthy living vegetation strongly absorbs red reflectance and strongly reflects NIR.

3.2. MOD 16 evapotranspiration data

We used the MOD16 global evapotranspiration product at 0.05° spatial resolution and 16-day temporal resolution. The ET algorithm is based on the Penman-Monteith equation (Monteith, 1965, Mu et al., 2011). We accumulated the global evapotranspiration starting in January and July to account for the different growing seasons in the Northern and Southern Hemisphere.

3.3. Market data

The data used in this paper is from a continuously updated price database comprised of food prices from 232 markets in 39 countries, collected by the FAO and the US Agency for International Development's Famine Early Warning Systems Network (FEWS NET). The data is available from the FAO at the Global Information and Early Warning System (GIEWS) website: http://www.fao.org/giews/pricetool/. We selected all the available markets that are located on the African continent. We have a total of 933 time series with market price information for 51 different products ranging from fuel to cattle to grain crops. The most common commodities were maize (137 price series), sorghum (118 price series), rice (102 price series), millet (56 price series), beans (43 price series) and cowpea (40 price series).The monthly time series differ in length with some starting as early as 1997 while others started in 2009. For time series with long data, only the period from 2000 was analyzed, and for those beginning after 2000, we report only the stations with at least four years of continuous data. We created a 0.5° buffer around the markets and determined the percentage of cropland within these buffers.

4. Methods

4.1. Land surface phenology metrics

To extract the peak height in NDVI and the peak timing based on accumulated evapotranspiration, we fit quadratic models with accumulated evapotranspiration based on MOD16 as the independent variable and NDVI as the dependent variable (Figure 1). We have used these models before based on Accumulated Growing Degree Days (AGDD) in other parts of the world (de Beurs and Henebry, 2008a,b, 2010). We have demonstrated recently that models based on moisture variables resulted in higher R^2 values in large areas around the world including Africa (Brown et al., 2012). In that paper we based our analysis on AVHRR data and we extend that work here with MODIS data. We fit the model two times for each pixel and year, once with data beginning in January and ending in December, and once with data beginning in July and ending in June. Figure 1 provides an example of the quadratic models for a market in Niger. For each 0.05° we calculate the NDVI peak height and the amount of accumulated evapotranspiration necessary to reach this peak NDVI. We also derive the day of the year for which the peak NDVI is reached.

4.2. Market analysis

Market prices are provided in monthly time series. Most market price series show a steady increase between 2000 and 2011 as a result of inflation and changes in world market prices. In this study we are not interested in the trend in these time series but rather in its seasonality. We calculate in which month, on average, the maximum price occurred, and we calculate the seasonal price spread. In addition, we calculate the difference between the maximum market price in each year and the minimum market price in the preceding eight months (Figure 2).

We apply Spearman's rank correlation to calculate the correlation between these price differences and the peak height based on our land surface phenology metrics (Figure 3).

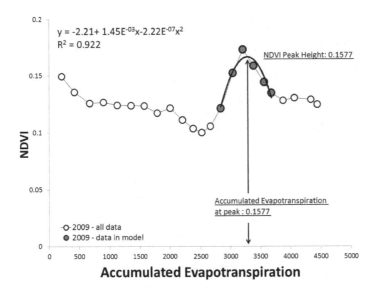

Figure 1. Example of a quadratic model fit for a station as far north as Ouallam, Niger. The example is created based on data from 2009. The NDVI peak height and the accumulated evapotranspiration at peak are calculated based on the quadratic regression model.

5. Results

5.1 Seasonal timing of vegetation and prices

The timing of the annual price maximum reveals a basic north-south pattern similar to the timing of the peak of the growing season observed by the land surface phenology models (Figure 4). The highest prices in West Africa occur in June and July in the far west, and in August and September in the central region. In East Africa the peak times occur in October and November, although a fair bit of variability can be observed. Southeastern Africa reveals highest prices in the months December through March. Figure 4 reveals that a great number of price time series peak around or slightly before the time that the vegetation as observed by satellite data peaks. There are a fair number of outliers, which are likely the result of non-environmental factors weighing more strongly on the price time series.

Figure 5 shows the seasonal price spread. The spread is highest for southeastern Africa and lowest for Western Africa. A low seasonal price spread indicates a certain amount of predictability about when prices may peak during the year.

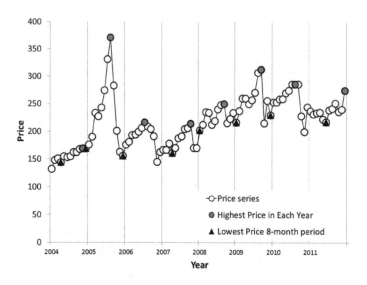

Figure 2. Example of a price time series of millet prices in Ouallam, Niger. For each year we determine the highest price (in grey dots). We also determine the difference between the highest price and the lowest price in the preceding 8-month period (shown in black triangles). Based on the highest prices we calculate in which month on average the highest price is found.

5.2. Correlation between vegetation peak height and price increases

We found that the significance of the correlation between price increases and NDVI peaks in the surrounding areas differed strongly by product. For example, we found that only 13% of the rice price series correlated significantly ($p < 0.1$) with NDVI peaks, while 35% of the cowpea price time series correlated strongly with NDVI peaks in the surroundings (Table 1).

Crop	Number of markets	% of sign. correlations
Cowpea	26	35%
Sorghum	76	32%
Beans	30	23%
Millet	41	22%
Maize	89	22%
Rice	46	13%

Table 1. Percent of significant correlation between market price increases and NDVI at peak by crop type.

Table 2 provides the 10 countries for which we have the largest number of market price series. We found a fair bit of variability in the number of significant correlations by country. For

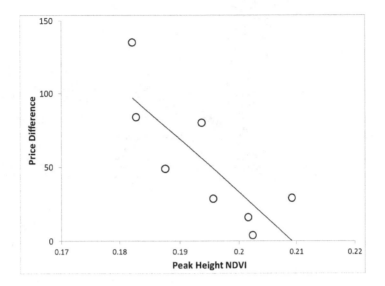

Figure 3. Example of correlation between the peak height of NDVI (Figure 1) and the price difference calculated based on market prices (Figure 2) for Ouallam, Niger. Each dot indicates a peak height x price difference combination for one year. We are showing the years 2004 – 2011. The final reported rank correlation for this area was -0.56, with a p-value of 0.037.

example, Nigeria, Somalia and Niger all revealed significant correlations between price increases and NDVI peak height for a large percentage of their markets (50%, 42% and 33%, respectively), while only 11% of the markets in Burundi revealed significant correlations.

Country	Number of markets	% of sign. correlations
Nigeria	22	50%
Somalia	45	42%
Niger	24	33%
Kenya	18	22%
Tanzania	28	18%
Mali	22	18%
Mozambique	22	14%
Burkina Faso	15	13%
Uganda	17	12%
Burundi	18	11%

Table 2. Percent of significant correlation between market price increases and NDVI at peak by country.

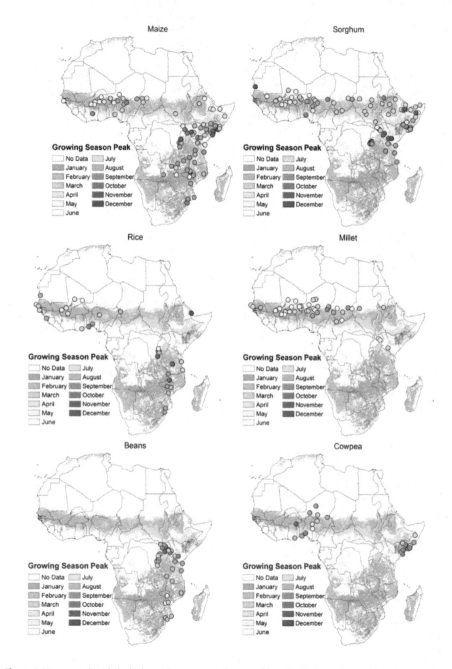

Figure 4. Mean month with the highest prices corresponds reasonably well with the peak timing based on NDVI.

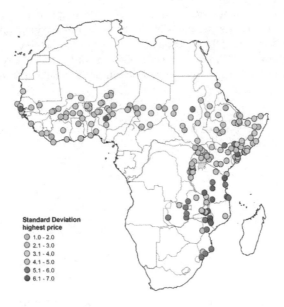

Figure 5. High seasonal price spread indicates more variability in the timing of the price peaks. Low price spread indicates that the annual maximum prices are found approximately in the same month every year.

5.3. Correlation between the amount of accumulated evapotranspiration at peak vegetation and prices

When we calculated the correlation between the peak timing in accumulated evapotranspiration and actual price values at their annual peak, we found a high number of markets with significant correlation, especially for millet. Beans revealed the lowest number of markets with significant correlation (16%). When we investigate the correlations by country, we find the largest number of significant correlations in Mali (Table 4), where 48% of the market price series correlate significantly with the peak timing in evapotranspiration units. Kenya and Burkina Faso also show a large number of significant correlations (45% and 40%). The lowest number of significant correlations was found in Tanzania, where only 15% of market price series correlated significantly with the amount of accumulated evapotranspiration at peak timing.

6. Discussion

In Sub-Saharan Africa, rice is the dominant commodity that appears to reflect changes in world prices (Kelly et al., 2008). Consequently, it is profitable to import rice, while arbitrage opportunities for grain trade are significantly lower. Further, several countries (many in West Africa) consume rice, but need to import rice to meet domestic consumption needs. Thus, our lack of correlation between rice price series and NDVI in the surrounding regions appears reasonable (Table 1 and 3). Millet is one of the products that can be grown in semi-arid zones and is most widely available (Brown, 2008). The ability of millet to grow in semi-arid zones likely results in the higher number of markets that reveal significant correlations between prices and evapotranspiration (Table 3).

Crop	Number of markets	% of sign. correlations
Millet	46	35%
Sorghum	84	32%
Maize	112	31%
Rice	54	30%
Cowpea	26	23%
Beans	36	16%

Table 3. Percent of significant correlation between market prices and amount of evapotranspiration at peak by crop type.

Eight of the countries investigated are classified as having low food security; Nigeria and Burkina Faso are classified as middle food security (Yu et al., 2010). All have unfavorable climates for agriculture, except for Uganda and Burundi, which also show the lowest amount of correlation between NDVI peak height and price increases (Table 2). It appears that price changes in these two countries are less affected by weather conditions. More than 70% of cereal needs are generally met through domestic production in Mali, Burkina Faso and Niger (Kelly et al., 2008). As a result, food prices tend to rise as a result of production short falls. We found that prices in these countries revealed very high correlation with peak height NDVI (Niger, Table 2) and amount of accumulated evapotranspiration at peak (Mali and Burkina Faso, Table 4).

Country	Number of Markets	% of sign. correlations
Mali	23	48%
Kenya	22	45%
Burkina Faso	15	40%
Nigeria	25	36%
Uganda	22	32%
Somalia	49	29%
Niger	24	25%
Mozambique	23	22%
Burundi	22	18%
Tanzania	34	15%

Table 4. Percent of significant correlation between market prices and amount of evapotranspiration at peak by country.

7. Conclusions

Preliminary conclusion based on this analysis is that we can get a better understanding of where satellite data could aid in the prediction of local market prices. For example, prices may be driven by different factors in Uganda and Burundi than in Niger and Nigeria. If we combine our knowledge of the effect of large scale climate oscillations on the land surface phenology (Brown et al., 2010) with the links between price time series and land surface phenology, we may be able to get an understanding of where we could predict local prices. Satellite data may be most effective in predicting local prices in poorly integrated markets. Poorly integrated markets are often 'thin' markets that have too little supply during times of high demand, such as right before harvest, resulting in food shortages. Seasonal price changes may reflect changes in production; in good years infrastructure and trade contraints reduce the ability of traders to move excess grain out of the area, and in poor years food is not moved into the area.

Author details

K. M. de Beurs[1] and M. E. Brown[2]

1 Department of Geography and Environmental Sustainability, The University of Oklahoma, Norman, OK, USA

2 Biospheric Sciences Branch, Code 614.4, NASA Goddard Space Flight Center, Greenbelt, MD, USA

References

[1] Alderman, H., Shively, G.E. (1996). Economic Reform and Food Prices: Evidence from Markets in Ghana. World Development 24, 3, 521-534

[2] Asante, E., Asuming-Brempong, S., Bruce, P.A. (1989). Ghana Grain Marketing Study. Ghana Institute of Management and Public Adminstiration and the World Bank, Accra and Washington DC

[3] Barrett, C. (2008). Spatial Market Integration. In: New Palgrave Dictionary of Economics, 2nd Edition.

[4] Brown, M.E. (2008). Famine Early Warning Systems and Remote Sensing Data. Springer Verlag, Heidelberg.

[5] Brown, M.E., de Beurs, K.M., Vrieling, A. (2010). The response of African land surface phenology to large scale climate oscillations. Remote Sensing of Environment, 114, 10, 2286-2296. doi:10.1016/j.rse.2010.05.005

[6] Brown, M.E., de Beurs, K.M. (2008). Evaluation of multi-sensor semi-arid crop season parameters based on NDVI and rainfall. Remote Sensing of Environment, 112, 5, 2261-2271. doi:10.1016/j.rse.2007.10.008

[7] Brown, M.E., de Beurs, K.M., Marshall, M. (2012). Global phenological response to climate change in crop areas using satellite remote sensing of vegetation, humidity and temperature over 26 years. Remote Sensing of Environment, 126, 16, 174-183. doi:10.1016/j.rse.2012.08.009

[8] Brown, M.E., Pinzon, J.E., Didan, K., Morisette, J.T., Tucker, C.J. (2006). Evaluation of the consistency of long-term NDVI time series derived from AVHRR, SPOT-vegetation, SeaWiFS, MODIS, and Landsat ETM+ sensors. IEEE Transactions of Geoscience and Remote Sensing, 44, 7,1787-1793.

[9] de Beurs, K.M., Henebry, G.M. (2004). Land surface phenology, climatic variation, and institutional change: analyzing agricultural land cover change in Kazakhstan. Remote Sensing of Environment, 89,4, 497-509. doi:410.1016/j.rse.2003.1011.1006

[10] de Beurs, K.M., Henebry, G.M. (2008a). Northern Annular Mode effects on the land surface phenologies of Northern Eurasia. Journal of Climate, 21, 4257-4279.

[11] de Beurs, K.M., Henebry, G.M. (2008b). War, drought, and phenology: Changes in the land surface phenology of Afghanistan since 1982. Journal of Land Use Science, 3, 2-3, 95-111.

[12] de Beurs, K.M., Henebry, G.M. (2010). A land surface phenology assessment of the northern polar regions using MODIS reflectance time series. Canadian Journal of Remote Sensing 36:S87-S110.

[13] Fafchamps, M. (1992). Cash crop production, food price volatility, and rural market integration in the third world. American Journal of Agricultural Economics, 72, 1, 90-99.

[14] Fafchamps, M. (2004). Market Institutions in Sub-Saharan Africa: Theory and Evidence. The MIT Press, Cambridge, MA

[15] Funk, C., Budde, M. (2009). Phenologically-tuned MODIS NDVI-based production anomaly estimates for Zimbabwe. Remote Sensing of Environment, 113,1, 115–125.

[16] Funk, C., Brown, M.E. (2009). Declining global per capita agricultural production and warming oceans threaten food security. Food Security, 1, 3, 271-289. doi:10.1007/s12571-009-0026-y

[17] IAASTD (2008) International Assessment of Agricultural Knowledge, Science and Technology for Development. Island Press, London

[18] Jennings, S., Magrath, J. (2009). What Happened to the Seasons? Seasonality Revisited. Oxford, England: Institute of Development Studies. A paper for the Future Agricultures Consortium International Conference on Seasonality, July 2009.

[19] Karnieli, A., Agam, N., Pinker, R.T., Anderson, M., Imhoff, M.L., Gutman, G.G., Panov, N., Goldberg, A. (2010). Use of NDVI and Land Surface Temperature for drought assessment: merits and limitations. Journal of Climate, 23, 3,618-633. doi: 10.1175/2009jcli2900.1

[20] Kelly, V., Dembele, N.N., Staatz, J. (2008). Potential food security impacts of rising commodity prices in the Sahel: 2008-2009. Famine Early Warning Systems Network (FEWS NET), USAID and Michigan State University.

[21] Kelly, V., Diagana, B., Reardon, T., Gaye, M., Crawford, E. (1996). Cash crop and food grain productivity in Senegal: Historical view, new survey evidence and policy implications. Productive sector growth and environmental division. U.S. Agency for International Development,.

[22] Lieth, H. (1974). Phenology and seasonality modeling. Springer-Verlag, New York, New York, USA.

[23] Lutz, C., Van Tilburg, A., van der Kamp, B. (1995). The process of short- and long-term price integration in the Benin maize markets. European Review of Agricultural Economics, 22, 191-212.

[24] Monteith, J.L. (1965). Evaporation and environment. In: Fogg GE (ed) Symposium of the Society for Experimental Biology, The State and Movement of Water in Living Organisms. Academic Press, New York,

[25] Mu, Q., Zhao, M., Running, S.W. (2011). Improvements to a MODIS global terrestrial evapotranspiration algorithm. Remote Sensing of Environment, 115, 1781-1800.

[26] Myneni, R.B., Keeling, C.D., Tucker, C.J., Asrar, G., Nemani, R.R. (1997). Increased plant growth in the Northern high latitudes from 1981 to 1991. Nature, 386, 698-702.

[27] Nemani, R.R., Keeling, C.D., Hashimoto, H., Jolly, W.M., Piper, S.C., Tucker, C.J., Myneni, R.B., Running, S.W. (2003). Climate-driven increases in global terrestrial net primary production from 1982 to 1999. Science, 300, 5625,1560-1563.

[28] Schaaf, C.B., Gao, F., Strahler, A.H., Lucht, W., Li, X.W., Tsang, T., Strugnell, N.C., Zhang, X.Y., Jin, Y.F., Muller, J.P., Lewis, P., Barnsley, M., Hobson, P., Disney, M., Roberts, G., Dunderdale, M., Doll, C., d'Entremont, R.P., Hu, B.X., Liang, S.L., Privette, J.L., Roy, D. (2002). First operational BRDF, albedo nadir reflectance products from MODIS. Remote Sensing of Environment, 83, 1-2,135-148. doi:10.1016/s0034-4257(02)00091-3.

[29] Slayback, D.A., Pinzon, J.E., Los, S.O., Tucker, C.J. (2003). Northern hemisphere photosynthetic trends 1982-99. Global Change Biology, 9, 1, 1-15.

[30] Tschirley, D., Weber, M.T. (1994). Food security strategies under extremely adverse conditions: The determinants of household income and consumption in rural Mozambique. World Development, 22, 2,159-173.

[31] Tucker, C.J., Pinzon, J.E., Brown, M.E., Slayback, D., Pak, E.W., Mahoney, R., Vermote, E., Saleous, N. 2005. An extended AVHRR 8-km NDVI data set compatible with MODIS and SPOT Vegetation NDVI Data. International Journal of Remote Sensing 26, 20, 4485-4498.

[32] Verdin, J., Klaver, R. (2002). Grid cell based crop water accounting for the Famine Early Warning System. Hydrological Processes, 16,1617-1630.

[33] Vrieling, A., de Beurs, K.M., Brown, M.E. (2008). Recent trends in agricultural production of Africa based on AVHRR NDVI time series. In: SPIE Europe Security + Defense, Cardiff, UK, September 15-18, 2008 2008.

[34] White, M.A., de Beurs, K.M., Didan, K., Inouye, D.W., Richardson, A.D., Jensen, O.P., O'Keefe, J., Zhang, G., Nemani, R.R., van Leeuwen, W.J.D, Brown, J.F., de Wit, A., Schaepman, M., Lin, X., Dettinger, M., Bailey, A.S., Kimball, J.S., Schwartz, M.D., Baldocchi, D.D., Lee, J.T., Lauenroth, W.K. (2009). Intercomparison, interpretation, and assessment of spring phenology in North America estimated from remote sensing for 1982-2006. Global Change Biology, 15, 10, 2335-2359.

[35] Yu, B., You, L., Fan, S. (2010). Toward a typology of food security in developing countries. IFPRI Discussion paper 00945.

[36] Zant, W. (2013). How Is the globalization of food markets progressing? Market integration and transaction costs in subsistence economies. World Bank Economic Review, 27, 1, 28-54. doi:10.1093/wber/lhs017.

Permissions

The contributors of this book come from diverse backgrounds, making this book a truly international effort. This book will bring forth new frontiers with its revolutionizing research information and detailed analysis of the nascent developments around the world.

We would like to thank Dr. Aondover Tarhule, for lending his expertise to make the book truly unique. He has played a crucial role in the development of this book. Without his invaluable contribution this book wouldn't have been possible. He has made vital efforts to compile up to date information on the varied aspects of this subject to make this book a valuable addition to the collection of many professionals and students.

This book was conceptualized with the vision of imparting up-to-date information and advanced data in this field. To ensure the same, a matchless editorial board was set up. Every individual on the board went through rigorous rounds of assessment to prove their worth. After which they invested a large part of their time researching and compiling the most relevant data for our readers. Conferences and sessions were held from time to time between the editorial board and the contributing authors to present the data in the most comprehensible form. The editorial team has worked tirelessly to provide valuable and valid information to help people across the globe.

Every chapter published in this book has been scrutinized by our experts. Their significance has been extensively debated. The topics covered herein carry significant findings which will fuel the growth of the discipline. They may even be implemented as practical applications or may be referred to as a beginning point for another development. Chapters in this book were first published by InTech; hereby published with permission under the Creative Commons Attribution License or equivalent.

The editorial board has been involved in producing this book since its inception. They have spent rigorous hours researching and exploring the diverse topics which have resulted in the successful publishing of this book. They have passed on their knowledge of decades through this book. To expedite this challenging task, the publisher supported the team at every step. A small team of assistant editors was also appointed to further simplify the editing procedure and attain best results for the readers.

Our editorial team has been hand-picked from every corner of the world. Their multi-ethnicity adds dynamic inputs to the discussions which result in innovative outcomes. These outcomes are then further discussed with the researchers and contributors who give their valuable feedback and opinion regarding the same. The feedback is then collaborated with the researches and they are edited in a comprehensive manner to aid the understanding of the subject.

Apart from the editorial board, the designing team has also invested a significant amount of their time in understanding the subject and creating the most relevant covers. They scrutinized every image to scout for the most suitable representation of the subject and create an appropriate cover for the book.

The publishing team has been involved in this book since its early stages. They were actively engaged in every process, be it collecting the data, connecting with the contributors or procuring relevant information. The team has been an ardent support to the editorial, designing and production team. Their endless efforts to recruit the best for this project, has resulted in the accomplishment of this book. They are a veteran in the field of academics and their pool of knowledge is as vast as their experience in printing. Their expertise and guidance has proved useful at every step. Their uncompromising quality standards have made this book an exceptional effort. Their encouragement from time to time has been an inspiration for everyone.

The publisher and the editorial board hope that this book will prove to be a valuable piece of knowledge for researchers, students, practitioners and scholars across the globe.

List of Contributors

M. B. Sylla and J. S. Pal
Loyola Marymount University, Seaver College of Science and Engineering, Department of Civil Engineering and Environmental Science, Los Angeles, CA, USA

I. Diallo
Laboratory for Atmospheric and Ocean Physics - Simeon Fongang, Polytechnic School, University Cheikh Anta Diop, Dakar, Senegal

J.G. Grijsen
Independent Hydrology and IWRM Consultant, Arrowlake Road, Wimberley, Texas, USA

A. Tarhule
Department of Geography and Environmental Sustainability, University of Oklahoma, Norman, USA

C. Brown
Department of Civil and Environmental Engineering, University of Massachusetts, Amherst, USA

Y.B. Ghile
Woods Institute for the Environment, Stanford University, Stanford, USA

Ü. Taner
Dept. of Civil and Environmental Engineering, University of Massachusetts, Amherst, USA

A. Talbi-Jordan
The World Bank Middle East North Africa (MNSWA), Washington, USA

H. N. Doffou, A. Guero, R. Y. Dessouassi, S. Kone and B. Coulibaly
Niger Basin Authority/ Autorité du Bassin du Niger (ABN), Niamey, Niger

N. Harshadeep
The World Bank, Africa Region, Washington DC, USA

M.G. Ogurtsov
Ioffe PhTI, St. Petersburg, Russia, Central Astronomical Observatory at Pulkovo, S. Petersburg, Russia

M. Lindholm and R. Jalkanen
Finnish Forest Research Institute, Rovaniemi, Finland

Marcela Hebe González
Department of Atmospheric and Oceanic Science - FCEN-University of Buenos Aires, Research Center of Ocean and Atmosphere – CONICET/UBA; UMI-IFAECI/CNRS, CIMA - 2° piso, Pabellón II, Ciudad Universitaria, Ciudad Autónoma de Buenos Aires, Argentina

Shahab Araghinejad and Ehsan Meidani
College of Agricultural Technology and Science, University of Tehran, Karaj, Iran

Renee A. McPherson
Department of Geography and Environmental Sustainability, University of Oklahoma, Norman, Oklahoma, USA

K. M. de Beurs
Department of Geography and Environmental Sustainability, The University of Oklahoma, Norman, OK, USA

M. E. Brown
Biospheric Sciences Branch, Code 614.4, NASA Goddard Space Flight Center, Greenbelt, MD, USA

Printed in the USA
CPSIA information can be obtained
at www.ICGtesting.com
JSHW011405221024
72173JS00003B/425

9 781632 391162